Academia and Industrial Pilot Plant Operations and Safety

ACS SYMPOSIUM SERIES **1163**

Academia and Industrial Pilot Plant Operations and Safety

Mary K. Moore, Editor
Eastman Chemical Company
Kingsport, Tennessee

Elmer B. Ledesma, Editor
University of St. Thomas
Houston, Texas

Sponsored by the
ACS Division of Industrial & Engineering Chemistry

American Chemical Society, Washington, DC

Distributed in print by Oxford University Press

Library of Congress Cataloging-in-Publication Data

Academia and industrial pilot plant operations and safety / Mary K. Moore, editor, Eastman Chemical Company, Kingsport, Tennessee, Elmer B. Ledesma, editor, University of St. Thomas, Houston, Texas ; sponsored by the ACS Division of Industrial & Engineering Chemistry.
    pages cm. -- (ACS symposium series ; 1163)
  Includes bibliographical references and index.
  ISBN 978-0-8412-2960-0
  1. Chemical plants--Pilot plants--Safety measures. 2. Chemical laboratories--Safety measures. I. Moore, Mary K. (Chemist) II. Ledesma, Elmer B. III. American Chemical Society. Division of Industrial and Engineering Chemistry. IV. Title: Industrial pilot plant operations and safety.
  TP155.5.A23 2014
  660--dc23
2014024916

Foreword

The ACS Symposium Series was first published in 1974 to provide a mechanism for publishing symposia quickly in book form. The purpose of the series is to publish timely, comprehensive books developed from the ACS sponsored symposia based on current scientific research. Occasionally, books are developed from symposia sponsored by other organizations when the topic is of keen interest to the chemistry audience.

Before agreeing to publish a book, the proposed table of contents is reviewed for appropriate and comprehensive coverage and for interest to the audience. Some papers may be excluded to better focus the book; others may be added to provide comprehensiveness. When appropriate, overview or introductory chapters are added. Drafts of chapters are peer-reviewed prior to final acceptance or rejection, and manuscripts are prepared in camera-ready format.

As a rule, only original research papers and original review papers are included in the volumes. Verbatim reproductions of previous published papers are not accepted.

ACS Books Department

Contents

Indexes

Preface

This symposium series volume was developed in order to share papers presented at the 245th ACS National Meeting, held April 7–12, 2013 in New Orleans, Louisiana. We believe this a first publication in this general area. The Industrial and Chemical Engineering, Applied Chemical Technology Subdivision hosted a one-day symposium for industry and academic researchers to present and share their work and best practices on operations and safety in pilot plant environments.

Designing and building "pilot plants" can be a big challenge, both to assure proper operations and to run the plant safely. Regardless if you are designing a plant in academia or industry, the safe operations of the pilot plant should remain a priority. One concern in building a pilot plant is the available real-estate. When designing and building pilot plants, several processes are considered such as chemical separation, chemical reactions, waste generation, process controls, etc. Plant designers and contractors also need to consider ease of operation and maintenance, pinch-points, sampling, etc.

In contrast to workers in large-scale facilities, pilot plant workers in both industry and academia are not only exposed to relatively small amounts of hazardous substances, they are exposed to such materials in a smaller setting and thus to potentially greater concentrations. We feel, therefore, that an opportunity existed to bring together workers and researchers in both industrial and academic pilot plant environments to present and share their best practices as regards operations and safety.

On the cover: Picture supplied by Eastman Chemical Company.

Mary K. Moore
Eastman Chemical Company, Eastman Research Division
P.O. Box 1972
Kingsport, TN 37662
(423) 229-1911 (telephone)
mkmoore123@gmail.com (e-mail)

The views presented here do not necessarily represent the views of my employer, and I do not have the endorsement of, or authority to speak on behalf of, my employer.

Elmer B. Ledesma
University of St. Thomas, Department of Chemistry and Physics
Houston, TX 77006
(713)-831-7810 (telephone)
ledesme@stthom.edu (e-mail)

Chapter 1

Academia and Industrial Pilot Plant Operations and Safety

Joel J. Justin*

TTM Division, Delta College, University Center, Michigan 48710, United States
***E-mail: joeljustin@delta.edu**

"Academia and Industrial Pilot Plant Operations and Safety" is written for the college or university environment in which students are developing the skills necessary to function within the process industry whether as an operator, researcher or design engineer. Proper management, eclectic exercises, training and hazard awareness are key parts of developing a safe and progressive learning environment. A clear and logical progression of necessary skills are essential to the development of responsible and safe process operations personnel. The student learner should be challenged on a daily basis and a dynamic approach to learning should be exercised by the instructor. This chapter will stimulate ideas while highlighting time tested techniques and methods currently being used.

The operation of a pilot plant in industry provides a necessary scaleup from the benchtop at an intermediate step before beginning full scale production. In most cases, the development of new compounds and their associated reactions will take place in 1-2 liter vessels while under initial research and development. Pilot plant vessels range from 100-500 liters and full scale production vessels are typically 20-40k liters. Pilot plant facilities operated within industry are often dealing with unknown hazards and materials that have not been previously reacted at this scale and often before any long term EH&S data has been compiled. Researchers are highly skilled and every precaution is taken to ensure that employee exposure and and any negative environmental impact are reduced or eliminated. Careful record keeping and strict GLP (good laboratory procedures) are followed. High Tech facilities utilize "soft" tools such as PHA (Process Hazard Analysis), PSM (Process Safety Management) and CHP (Chemical Hygeine Plans) to best prepare for any unforeseen – yet predictable occurrence. Focus is on repeatability, predictable and incremental changes, and detailed analysis before any deviation is allowed. While very useful, not every project moves forward with a complete and accurate P&ID (Piping & Instrument Diagram) to aid with their PHA efforts.

Researchers are experienced, most often have worked together in similar environments and have open and functional communication between all team members. Every precaution is considered during pre-task planning in a concerted effort to eliminate any "suprises" that may occur once an experiment begins. Still, accidents and exposures do occur within the pilot plant industry. If one scans the statistists published on the OSHA.gov website some patterns emerge. These include incidents involving poor or inadequate instrumentation, poor or little control of unwanted combustion within confined spaces. This can best be completed by 1 or both of the following:

1) **Nitrogen inerting**-while Nitrogen introduces an additional asphyxiation hazard to the laboratory environment, it has a long history of enhancing safety through the reduction of explosion hazards. Since the use of only non-sparking equipment is very improbable, the removal of oxygen of confined spaces is then a very viable method to reduce/eliminate explosion hazards. Students should be instructed in the hazards associated with nitrogen including:

   - Odorless and colorless gas (poor warning properties).
   - Commonly used throughout the chemical industry.
   - Operations personnel often become complacent in working in areas where nitrogen gas is in use.

2) **Bonding and grounding of equipment**-since the transfer of almost every solid, liquid, gas or slurry creates a static charge, the bonding and grounding of process equipment is essential to prevent unwanted discharge of electrical potential – often the root cause of explosions within pilot facilities. The proper maintenance of this system is key to ensure electrical continuity as equipment and lines are disassembled for inspections and cleanout. This requires well trained operators and a re-commissioning process for putting equipment back online. Often this means the use of a electrician's multimeter before equipment can be considered "bonded".

While pilot plant operations within industrial settings involve hazards associated with reactions and mixtures that are relatively "untested", the hazards associated with pilot plants within academic settings involve potential hazards associated with the lack of knowledge of the students. Unlike the trained professionals that inhabit the real world counterpart, student learners lack the same experience and knowledge. Therefore careful planning, preparedness and open communication must be emphasized in order to minimize hazard.

Ideally, a 50/50 mixture of classroom lecture and hands on laboratory exercises should be considered in order to ensure an optimum learning environment. Through experience we have learned that a 4 hour period is on the average the longest that a group of students should be subjected to. After 4 hours, focus and attention tends to decline and the learning curve begins to decline. Additionally, within the laboratory the incidence of exposure to hazards would also increase.

The goal of the classroom lecture is to keep pace with the current hands on exercise. Incrementally, new skills are added at a pace determined by student learning progress and that of an experienced teacher. Common themes that must be emphasized and supported from the very first day include safety, teamwork, communication, troubleshooting skills, documentation, and record keeping. Every attempt should be made to emulate the standards and culture of the industry you are preparing the students for when seeking employment. For technical schools and colleges this usually involves local industry. For universities, a broader based approach may best serve the student learner.

Industrial facilities within your geographic area most likely will be very helpful, if not enthusiastic in helping your staff with setting up program parameters. The involvement of these companies as members of an "advisory board" not only provides real time input, but also develops a liason for spare parts, jobs for your graduates and often a pool of well qualified guest speakers or adjunct instructors. The utilization of SME's (Subject Matter Experts) when developing your program, or when making changes to your program, is the best way to stay current with area hiring needs and may provide a cost effective way for handling expensive repairs as well as reduction in the financial impact of capital projects. If asked, most area companies may willingly loan skilled trades workers to help with your maintenance needs. For example: most well equipped pilot facilities have instrumentation and DCS systems that are far beyond the knowledge of a "general electrician". In these cases, access to a journeyman "instrument technician" is invaluable. Since this trade is so specialized, they are most likely employed only by mid-sized to larger industrial facilities.

Therefore, the major difference between an industrial pilot plant and an academic pilot plant lies in the "unknown" factor involved. In industry this involves the hazards associated with unknown reactions, compounds/chemicals, and the pressures associated with deadlines and corporate culture. With pilot plants within academic settings, the unknown lies with the student learner.

Safety

Safety, in any setting is a mindset. Safety within a workgroup, is a culture. When workers are asked 2 simple questions, this is easily exhibited:

Q1: What percentage of workers in your plant comply with all safety standards and PPE requirements during the day shift when management is present?

Q2: What percentage of workers in your plant comply with all safety standards and PPE requirements during the off shift when there is no management present?

Employees will agree that there is a moderate to significant drop in safety compliance when workers are not being observed. It is important to note that the hazards in the workplace actually increase during the nighttime and afternoon shifts. Worker fatigue, lack of adequate lighting, fewer employees to share tasks, as well as many other factors are all contributors. <u>Therefore, one can easily extrapolate that as the potential for a hazard increases during the "off shifts", the actual compliance with safety standards and PPE requirements decreases.</u> Data supports that the likelihood of an accident to occur is much more likely during the night shift. Most famously are major catastrophes such as the Exxon Valdez, Chernobyl, Titanic, Piper Alpha Rig, and Three Mile Island disasters.

Therefore, it is extremely important that students learn safety from day one and that it is emphasized at all aspects of their learning process. Students must learn how to recognize hazards, this is done through training and the consistent application of all standards by their instructors. Students should also learn to intervene when fellow students are at risk due to improper PPE and/or compliance with laboratory rules and safety standards.

The culture of safety within a work environment is an accumulation of behaviors and the attitude of those workers. This may be affected in a positive or negative manner by the following factors:

1. Workplace Culture
2. Complacency
3. Training
4. Hazard awareness
5. Distractions/motivation
6. Focus
7. Communication

The implementation of BBS (Behavior Based Safety) has helped to revolutionize safety for a large portion of the chemical processing industry both within the United States and abroad. BBS also fits very well in the academic laboratory setting and helps the student to build the necessary safety awareness and attitude coveted by future employers. Simply put, BBS safety requires that the students and instructors:

1. Keep a positive environment
2. Celebrate the good points (large or small)
3. Set goals – success mapping
4. Utilize teamwork and positive intervention
5. Set the right example

6. Develop a safety "culture"
7. Develop "habit strength" behaviors towards safety performance

<u>"Everyone likes to hear they are doing a good job!"</u> Reinforcing behaviors POSITIVELY is the driving force to the success of BBS in the reduction of accidents in the workplace. While BBS provides the vehicle in which a successful safety culture thrives, several tools are also necessary and available to keep safety performance in the forefront:

Measure

Keep a spreadsheet of the team's safety performance history and <u>always make it visible to the students.</u> Use a formula of (# of students X laboratory hours X laboratory days X Classes being held) to get "Total Consecutive Student Laboratory Hours Without an Accident" Students will respond positively and take pride in these statistics.

Celebrate Success

Set safety goals and celebrate when they are reached. Students will associate safety performance with rewards.

Maintain a POSITIVE Environment

This helps to reduce the negative effects of the factors related to cultural and industry related influences.

Accountability

Ensure students are accountable for their own performance. This includes sign in/sign out sheets, attendance, safety compliance, record keeping and ability to follow established SOP (Standard Operating Procedure). It is suggested that each of these are measured, and become a part of each student's scoring rubric.

Involve Students in Decision Making

- Develop a discussion panel that encourages student input (safety, technology, budget).
  - Involve students in RCI's (Root Cause Investigations)
  - Encourage student participation in advisory panel discussions.
  - Require the writing of SOP (Standard Operating Procedure) as part of graded curriculum.

Stay Current with Industry/Regulations

Instructors should visit/tour area plants and discuss the needs of local industry. Training in EH&S (Employee Health and Safety) is readily available and state OHSA extensions offer free or relatively inexpensive training options.

Incorporate Industry SME's (Subject Matter Experts)

The involvement of area experts is invaluable to your program, your students and therefore the industry as a whole. Most area employers are very willing to assist you with your needs – including providing expert help (instruction, maintenance, inspections, advisory) whenever asked.

Preparing Students for Employment in Industry

The best way to satisfy employment needs, as well as to best prepare the student for real world employment, is to utilize all of the tools being used by area industry.

Good Laboratory Practices (GLP)

Proper documentation, log entrys and corrections as per industry standards. Once introduced, the student's laboratory writeups must be closely monitored so as to prepare them for proper documentation compliance once employed. This includes the use of black ink, mistake correction (single line, initialed and dated), not skipping data entry boxes, not erasing and the elimination of correction fluids (*1*).

Standard Operating Procedures (SOP)

Students must learn to interpret, adhere to and troubleshoot with SOP's. Students should learn to write, test, develop and publish SOP's. These include P&IDs (Process and Instrument Diagrams), startup and shutdown procedures (*2*).

BOLS (Breaks-Odors-Leaks-Spills)/LOPC (Loss of Primary Containment)

Forms obtained from local industry and used in the pilot facility will help students prepare for employer's expectations for environmental release reporting and participation with maintenance trends. Students have to identify, log in, clean up and report any spills resulting from Loss of Primary containment. A monthly review of these forms as a class helps to identify redundant issues, target PM (Preventative Maintenance) efforts and open communication between work groups.

Preventative Maintenance (PM)

Students should learn to perform all preventative maintenance to keep pilot plant in peak operating condition. Once PM interval spreadsheet is developed (with student input), students should be given responsibility. Tasks include lubrication, non destructive testing, line labeling, valve maintenance/packing, vibration testing, inventories and other tasks unique to your laboratory (*3*).

Troubleshooting Methodology

Troubleshooting is an essential competency to any employee in the chemical processing field. This is a valuable tool that will enable a student to approach an abnormal condition in a logical, organized and calm manner. The student must learn to recognize, investigate, prioritize and find a solution to process problems. Below is a suggested seven step process:

1. Identify the problem

   - The first step to solving any problem is to recognize that a problem exists, and then define it.

2. Identify any potential causes

   - Once a problem is found, this step is used to identify all potential sources/causes. It is important in this step to not omit any potential causes/sources of the problem step to ensure that all potential sources are considered.

3. Narrow the possibilities to the "likely" cause

   - Implementation of past experience with this equipment, worker's knowledge, conditions specific to this equipment will lead operator to the most likely cause of problem.

4. Draw preliminary conclusions

   - In this step a scenario should be developed that could reasonably explain how the likely cause in step 3 could result in the identified problem from step 1. This will lead personnel to consider all ancillary causations including those that may affect quality or safety.

5. Prove conclusions

   - In this step, the operator uses multiple process indicators (e.g., DCS (Distributed Control System) indications, field indications, sensory information, etc.) to verify that his or her conclusions are correct before taking action. This is the most important step in the methodology because it prevents taking an incorrect action. This step, more than any other, requires an in-depth understanding of the process operation and variable interactions

6. Implement corrective action(s)

   - Here, personnel will take action to correct the original problem and return the process to safe – or steady state condition.

7. Document

   - A Key step to operations excellence! Once the problem has been corrected, the last step is to document the problem, its solution and, most importantly, the troubleshooting process. Proper documentation prevents future occurences, streamlines your process and helps eliminate unnecessary downtime.

Pre-Task Analysis

A pre-task analysis is a formal way to ensure that a team is focused and best prepared to perform a task on a specific piece of equipment. Typicall in a written form, it is a tool to help a workgroup focus on the task at hand, identify hazards and communicate with one another. If properly implemented, the PTA will help a workgroup transition into a focused and coherent group. For example: discussing emergency shutdown procedures in a calm and deliberate setting will prove invaluable in times of emergency shutdown, often under duress and with limited time. Studies show that a PTA significantly improves a teams preparedness in times of emergency and in preparing for an upcoming task/job (*4*).

Laboratory Sign in/Sign out

In industry, all workers are required to diligently sign in and out of a building so as to be accounted for in the occurrence of a plant evacuation or assembly exercise. There should be no exception to this when it comes to the academic setting and will not only ensure accountability, but also help student to prepare for a position post graduation.

Tailgate Sessions

Each lab class should begin with a 5-15 minute tailgate session which helps the student transition mentally to being the best prepared student possible. In these brief instructor led sessions, the introductions of new skills, reinforcement of previously learned tasks and a reiteration of safety focus are recommended.

Safety Shower/Eye Bath

Students should learn how to test and take care of the safety shower and eye bath stations. This includes a weekly water clarity and temperature check. Area should be clean and clear of debris so there are no obstacles to impede a person with impaired vision attempting to get to the safety station. The eye cups should be in place at all times on the eye bath station so no foreign objects would additionally injure the person seeking help. As expected in industry, future operators should learn how to maintain this essential safety equipment (5).

Lock Out – Tag Out

The objective of the Lockout – Tagout process is the control of hazardous energy. All operating personnel must become familiar with the Lock Out/Tag Out standard. Whether as a Authorized or Affected worker, a functional understanding of this standard is key to a safe career as a process operator/laboratory operator. Students should learn to plan and perform Lock Out/Tag Out and learn how to return equipment that has been secured back into service (6).

Root Cause Investigations (RCI)

An invaluable tool implemented post-incident to help team to determine pathology of a safety or quality issue. It is recommended that instructor facilitate the procedure and solicit input from students representing the work team. Provides a lucid method of gathering ideas and forming a solution. Promotes buy-in from all team members and helps form a positive platform in which to encourage new ideas.

Personal Protective Equipment (PPE)

PPE standards must be developed to address all hazards that may be encounted within the laboratory environment. Additionally, students must be taught how to don, doff and properly clean and store PPE equipment. Included in the training should be a classroom portion to discuss theory and a practical application on the laboratory floor. It is suggested that students learn to complete everyday tasks (sampling, using hand tools, filling out runsheets, etc) while wearing PPE so as to understand the barriers/restrictions that can arise while in the Pilot Plant environment (7).

National Fire Protection Act (NFPA)

Participation in the labeling of process equipment as per the National Fire Protection Act will aid the student in recognizing and adhering to the recommended standards. Theory should be emphasized in the understanding of the color and numbering system as well as the icons for special hazards (8).

Sampling Safety

The safe sampling of process fluids, gasses and solids is a key part of a process operators position. Students in the pilot plant environment must learn how to sample safely, sanitarily, and utilize good GLP methods. A sampling procedure should be written to aid the student (figure 1).

An example of sampling procedure:

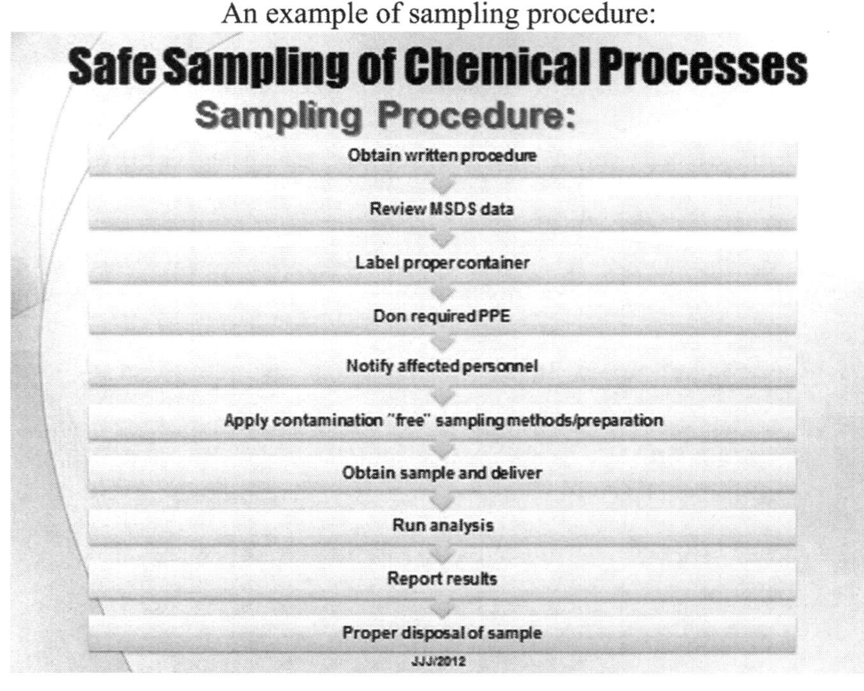

Figure 1. Field Sampling Method

The chart below can be utilized to help the student understand the matricies involved with the sampling of production streams within a chemical plant (figure 2).

Types of in situ sampling to be completed in laboratory:

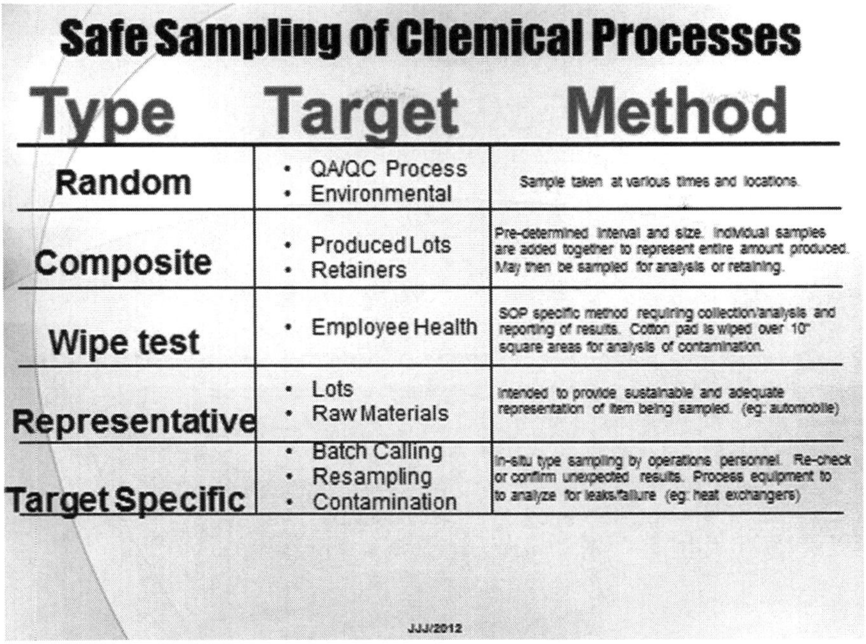

Safe Sampling of Chemical Processes

Type	Target	Method
Random	• QA/QC Process • Environmental	Sample taken at various times and locations.
Composite	• Produced Lots • Retainers	Pre-determined interval and size. Individual samples are added together to represent entire amount produced. May then be sampled for analysis or retaining.
Wipe test	• Employee Health	SOP specific method requiring collection/analysis and reporting of results. Cotton pad is wiped over 10" square areas for analysis of contamination.
Representative	• Lots • Raw Materials	Intended to provide sustainable and adequate representation of item being sampled. (eg: automobile)
Target Specific	• Batch Calling • Resampling • Contamination	In-situ type sampling by operations personnel. Re-check or confirm unexpected results. Process equipment to to analyze for leaks/failure (eg: heat exchangers)

JJJ/2012

Figure 2. Process Sampling Types

In addition, students can run analysis on a variety of process samples and depending on the equipment available to them. Some examples of "typical" analysis performed by process operators:

- Mass Spectrometry
- Gas Chromatography
- Liquid Chromatography
- Particle Size Distribution
- Percent Actives
- pH
- Titration
- Percent Solids
- Percent Moisture

Proper Pump Startup/Shutdown

The safe startup of process pumps is a skill set that needs to be taught by the instructor. Improper startup and shutdown is a major cause of early failure, environmental releases and chemical exposure to personnel throughout the process industry. If properly installed, a positive displacement pump should have a check valve installed on the discharge side. The check valve needs to be maintained in proper working condition. Additionally, any positive displacement pump should be equipped with a check valve on the discharge side in case of deadheading. All rotary or centrifugal pumps should also have a discharge side check valve installed. Students must be taught how to shut down a pump so as not to leave liquid filled. If done improperly, the pump will be a hazard to anyone performing maintenance due to the exposure to the chemical. In addition, liquid should not be left in contact with the seals for a long period of time and pressure can develop as a result of long term containment within the pump housing. To shut down the pump properly the following should be followed:

- Close the intake valve (this allows the pump to cycle out all of the liquid in the pump housing)
- Close the discharge valve (preventing any backwashing of liquids)
- Turn off the power

(Note: as an aid to help students remember, use – "Starve, Choke, Kill" – Figure 3)

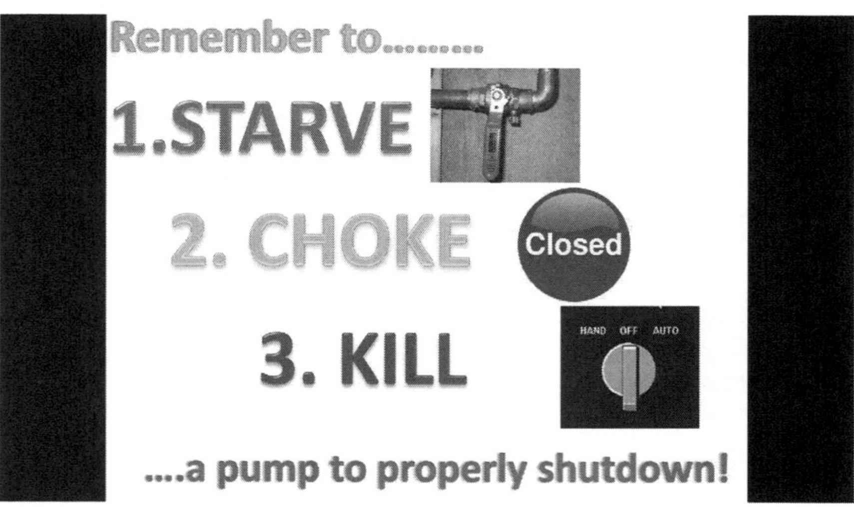

Figure 3. Safe Process Pump Shutdown

Startup of a pump requires the laboratory student to properly valve in the pump first while paying close attention to its flowpath. This may involve the shutting of valves to destinations not desireable. Students should learn that all lines in a process facility need to be terminated before energizing a pump such a plugs, caps, blind flanges or closed valves. It is imperative that students understand fully that once energized, THEY are responsible for that fluid transfer.

Proper Valve Operation

Laboratory students must learn how to properly open and close process valves. Each type of valve will be encounted in the process industry whether ball, gate, globe, butterfly, needle, diaphragm, plug or knife and a basic knowledge of each should be emphasized in the classroom. <u>It is essential that ball and plug valves be understood and the hazards associated with water hammer.</u> Since these two types only require a quarter turn of the handle to go from fully open to fully closed, the student must learn to hand actuate these valves carefully so as to reduce any surge of pressure or flow internally. *It should be emphasized that no liquids should be trapped between valves and left to reside in process equipment.*

Pipe Tracing

A pilot plant student who is training to be a process operator must be encouraged to learn how to trace the origin and insertion of process lines. In most positions within the process operator field, this is a key skill that they will be expected to master. The instructor can best prepare the student by insisting that each line in the laboratory be traced before any transfer occurs.

Piping Standards

The ability to assemble and perform maintenance on process piping is an important skill for today's process operator. Student should learn how to properly use hand tools to loosen or tighten pipe and pipe fittings. It is important to introduce the metallurgy involved with process piping and associated issues such as "galling" should be discussed. Below is a partial list of process piping used today:

- PVC
- Conduit
- Stainless Steel
- Fiberglass
- Carbon Steel
- Aluminum
- Copper
- Titanium
- Inconel
- Hasteloy
- Plastic Lined

A student should learn the types of ways that piping is joined and be able to recognize each:

- Screwed
- Bonded
- Glued
- Welded
- Flanged

Additionally a pilot plant student should know the types of fittings used in their process and how to install/maintain each:

- 45 degree elbow
- 90 degree elbow
- Pipe unions
- Pipe couplings
- Bushings
- Pipe plugs
- Pipe caps
- Orifice plates
- Blind flanges
- Pipe T
- 4 way connector

A keen ability to recognize piping schemes using a Piping and Instrument Diagram and then applying it to the process equipment is an essentials skill for a process operator.

Housekeeping

The proper maintenance and upkeep of a safe and environmentally sound laboratory involves the cleaning up of process equipment. Hydraulic fluid, oil, dust, process liquids or powders, grease, thread tape, spare parts – are all part of the hazards that can make an otherwise safe environment into an unsafe one.

Interval Lead Operator Roles

In laboratory teams within the pilot facility the lead role should be rotated to enable a 360 view of team dynamics. Students who experience this role are typically responsible for the division of tasks, all paperwork, GLP standards and the overall success/failure of the laboratory team. Employers in the chemical processing industry typically consider "leadership" skills as a key competency for employment. Coaching your student in this role will help to build their confidence and provide them with operational experience. Students should be taught how to listen, direct, intervene, lead and be accountable in guiding their laboratory team.

Teamwork Dynamics

Learning how to perform as part of a functional team is a key skillset looked for by today's employers. The values and culture of teamwork should be displayed by the instructor and learned by the student. Team building exercises, tailgate sessions, shared responsibility and inclusion are all tools that can help students to learn the fundamentals.

Rotation of Laboratory Teams

Once students are placed into teams, a rotation on a weekly basis will help students to progress through the facility in a logical manner. It is suggested that the laboratory be divided up into 4-5 unique exercise areas in which the teams will rotate through. A proper understanding of pre-task analysis, troubleshooting methodology and teamwork dynamics will greatly enhance the success of each individual team.

Below is a basic visual chart to illustrate this technique (Figure 4).

Chemical Process Lab Assignments

(Tue/Thu)	Process #1	Process #2	Process #3	Flex Group
Group A1				
Student 1	W4	W1	W2	W3
Student 2	W4	W1	W2	W3
Student 3	W4	W1	W2	W3
Student 4	W4	W1	W2	W3
Group A2				
Student 5	W2	W3	W4	W1
Student 6	W2	W3	W4	W1
Student 7	W2	W3	W4	W1
Student 8	W2	W3	W4	W1
Group A3				
Student 9	W3	W4	W1	W2
Student 10	W3	W4	W1	W2
Student 11	W3	W4	W1	W2
Student 12	W3	W4	W1	W2
Group A4				
Student 13	W1	W2	W3	W4
Student 14	W1	W2	W3	W4
Student 15	W1	W2	W3	W4
Student 16	W1	W2	W3	W4

*Flex Time = Project work/Cleanup/Assigned Tasks
*All students must sign in/sign out of lab
*Tailgate meeting 8:00 am sharp!

Figure 4. Student Rotation Schedule

Process Variables

The pilot plant environment is often a combination of "old technology" and "new technology" – as are many jobs within industry. Students should learn how to comfortably work with either. The understanding of the process variables should be taught in theory within the classroom and supported/experienced in a practical manner within the laboratory. These include:

- Pressure
- Flow
- Level
- Analytical
- Temperature

These process variables (PFLAT), must be understood and how they affect one another. The measurement of each, as used in the process is also a critical portion to becoming a competent chemical operator. Redundant instrumentation should be installed and students should comfortably be able to convert Celcius to Fahrenheit, Delta T, Delta P, PSIA to PSIG, Atmospheric Pressure and perform heat calculations.

Students should also become competent in calculating process signals whether in digital or analog format. This would include a comprehensive understanding of 3-15 psi (pneumatic signals) and 4-20 mA (electronic signals). For example: an entry level operator fully understands an analog variable that is at 50% capacity is seen by the process computer at 0.50, and has a pneumatic signal of 9 psi as well as an electronic signal of 12 mA. Below is a list of typical conversions used by students studying to become process operators:

Mass & Weight

- Density of water = 8.33 lb/gal = 62.5 lb/cu.ft.
- 1 ton = 2000 lb
- 1 kg = 2.2 lb
- 1 lb = 453.6 g

Temperature

- F to C: C = (f-32)/1.8
- C to F : F = (C*1.8)+32
- Absolute Zero = -273 degree C = -460 degree F

Pressure

- PSIA = PSIG + 14.7
- PSIG = PSIA – 14.7
- 1 PSI = 2.31' of water
- 1 PSI = 27.7 " of water
- 1 PSI = 2.04 " Hg
- 1 PSI = 51.7 mm Hg
- 1 PSI = 6.90 kPa

- 1' water = .433 PSI
- 1 ATM = 14.7 psi = 29.9 inches mercury = 760 mm mercury
- Total vacuum = 29.9 inches of mercury vacuum
- 1 ATM (sea level) = 14.7 psia or 29.9 Hg abs or 760 mm Hg abs

Volume

- 3.79L = 1 gal
- 1L = .264 gal
- $1 \text{ ft}^3 = 1728 \text{ in}^3$
- $1 \text{ ft}^3 = 7.48$ gal
- $1 \text{ ft}^3 = 28317$ cc

Viscosity

- Centipoises = Centistokes x SG

Metric – English

- 1kg = 2.2 lb
- 1 mile = 1.61 km
- 1" = 2.54 cm

Length & Area

- 1 mi = 5280'
- $1 \text{ ft2} = 144 \text{ in}^2$

Signal Conversions

- mA = 4-20
- PSI = 3-15
- mA to PSI : PSI = ((mA-4)/16)*12+3
- PSI to mA : mA =   ((PSI-3)/12)*16+4

Density & Specific Gravity

- Density (ρ) = mass/volume
- SG = $\dfrac{\text{Density of Material}}{\text{Density of Water}}$
- Water Density = 62.4 lb/ft^3
- Water Density = 1.0 g/ml

Flow

- FLOWmass = FLOWvolumetric X Density

Ideal Gas Law

- $\dfrac{P1\ V1}{T1} = \dfrac{P2\ V2}{T2}$

Heat Formulas

- 1 BTU = amount of heat required to raise 1 lb water 1 degree F
- $Q = M\ C\ \Delta T$
- $\mathbf{Q_H = Q_M\ C\ \Delta T}$ or $\mathbf{Q_H = Q_V\ \rho\ C\ \Delta T}$
- $Q = L_f\ M$
- $Q = L_v\ M$

- $Q_H = \underline{K\,A\,\Delta T}$ for heat conduction through single layer L
- $Q = \underline{K\,A\,t\,\Delta T}$ for heat conduction through single layer L

- $R = 1 / K$

- $Q_H = (A\,\Delta T)/R_T$ for multi-layer heat conduction

    Where $R_T = L_1 R_1 + L_2 R_2 + L_3 R_3 +$ ……………………

- $Q = UA\Delta T$ for heat exchangers, where U is the Overall Heat Transfer Coefficient

Level vs. Liquid Head Pressure

- $P = h \times 0.433 \times SG$

- $H = (P \times 2.31) / SG$

    - **P** = Liquid Head Pressure (psi) **H** = Liquid Head Height (ft) **SG** = Specific Gravity of Liquid

Rectangle or Square

- Area = l x w

- Perimeter = 2l x 2w

Triangle

- Area = ½ b x h

- Perimeter = a+b+c

Circle

- Area = pi x r^2

- Pi = 3.14

- Circumference = pi x d

Cylinder

- Volume = pi x r^2x h

Sphere

- Volume = 4/3 x pi x r^3

D/P Flow Measurement

$$\text{FLOWmeas} = \sqrt{\frac{\text{D/PMeas}}{\text{D/PMax}}} \times \text{FlowMax}$$

Piping and Instrument Diagram (P&ID)

The use of Piping and Instrument Diagrams to troubleshoot and locate process equipment is a basic skill of process operators and should be first learned in the pilot facility. Types of process lines, identifying equipment and field location involves classroom and laboratory engagement. Piping standards (ASME) should be reviewed and familiarized between staff and students. Students should also be familiar with Block Flow Diagrams (BFD), Process Flow Diagrams (PFD) and Isometric Drawings. The introduction of process drawings should begin within the classroom setting and be reinforced in the laboratory environment. The student should be competent in identifying devices contained within the drawings and be able to demonstrate wherer it is physically located in the field.

Communication

Learning to communicate effectively is a key skillset for anyone working in the chemical industry. A student should be introduced to, practice and be encouraged to become fluent in the many types of communication utilized by the chemical processing industry. The list may include:

- Verbal communication

    - 1 on 1
    - Radio
    - Telephone
    - PA

- Non verbal communication (hand signals)
- Process area signs
- Warning lights
- Painted warnings
- Warning sirens
- Written communication

    - Logbook
    - SOP
    - Written procedures
    - Hand notes
    - LO/TO
    - Email
    - Printed material

- Multi Media Communication

    - TV
    - Online delivered training
    - Communication boards/bulletins

Emergency Alerts/Shutdown

The practice of emergency alerts for evacuation and assembly can help laboratory personnel prepare for actual emergencies. When properly utilized, the pre-task analysis exercise will demonstrate that an emergency plan created by the team under "no duress" is an invaluable tool in the case of an emergency for securing process equipment in a safe manner.

Process Safety Management (PSM)

The validity and usefulness of this safety standard should be taught to your laboratory students as an early part of their learning experience. An effective technique is to implement a research project for each student in which they have to demonstrate an understanding of a particular standard to display comprehension. The PSM is a useful tool for the new and the seasoned employee within the chemical industry and the ability to access and understand the information is an important competency (9).

MSDS

Laboratory personnel should become familiar with accessing and understanding the MSDS system. Below are some helpful websites for the instructor:

https://www.osha.gov/dsg/hazcom/index.html
https://www.osha.gov/dte/outreach/intro_osha/intro_to_osha_guide.html

In addition to the teachable skills that your program should include, the laboratory should meet all EPA, OSHA and DEA requirements. These would include:

- Facility safety shower and eyebath
- MSDS for all materials
- Flammable Cabinet for all solvents/catalysts
- Interval sampling of air quality
- Proper waste disposal method (solid and sewer)
- Air quality permit (if applicable)

- Installation of safety devices and frangibles on all pressurized process equipment.
- Proper grounding of process equipment
- Switchroom that meets all applicable safety standards
- Electrical classification (if applicable)
- Easily accessible and appropriate PPE
- Application of Engineering Controls and Administrative controls as per OSHA standards.
- NFPA labeling of process vessels
- Labeling of process lines

Reaffirming Skills through Practical Application

Assignment of special projects will enhance the student's learning and support the theory that is delivered in the classroom setting.

Line and equipment labeling
Field sampling
Pipefitting
Line and equipment openings
Confined space entry
Lock out – Tag out
Pump disassembly
Scavenger hunts
Hand tool use
Blanking lines
Evacuating lines
Pressure checking lines
Field calculations
Filter changes
Fluid transfer
Drum/fiber pack filling
Hazard recognition

Measuring Student Competencies

It is essential that the practical compentency of your student in quantified. A matrix should be developed in which the student can be observed and measured in an "individual" basis. This is to identify any student who may have missed some practical skill development due to their participation as a member of a team in prior classroom settings. For example – see (Figure 5).

Laboratory Competencies Required for Process Operator

Passed

- Plan/prepare and perform a LO/TO for a process line opening.
- Safely demonstrate how to break and open a 4 bolt flange for gasket change on a process line.
- Safely demonstrate a technique to loosening a pipe union above operator's head height.
- Demonstrate the safe startup and shutdown of an in situ process pump.
- Display competency in hand tool use including standard and metric machined tools as well as tools intended for cylindrical pipe.
- Describe/demonstrate in detail the understanding of filter plugging, preparation and changing of process filters.
- Extrapolate flow measurement of process flow when given Delta T and coolant flow across a heat exchanger.
- Troubleshoot a faulty automatic valve.
- Demonstrate a bumpless transfer on a PID controller
- Troubleshoot a level device.

Instructors Signature:____

Figure 5. Measureable Student Competencies

APPENDIX

- ANSI – American National Standards Institute
- ASME – American Society of Mechanical Engineers
- BBS – Behavior Based Safety
- BOLS – Breaks Odors Leaks and Spills
- CHP – Chemical Hygeine Plan
- DCS – Distributed Control System
- EPA – Environmental Protection Agency
- GLP – Good Laboratory Practice
- LOPC – Loss of Primary Containment
- MSDS – Material Safety Data Sheet
- NFPA – National Fire Protection Act
- OSHA – Occupational Safety and Health Administration
- P&ID – Piping and Instrument Diagram
- PHA – Process Hazard Analysis
- PM – Preventative Maintenance
- PPE – Personal Protective Equipment
- PSIA – Pounds per Square Inch Actual
- PSIG – Pounds per Square Inch Guage
- PSM – Process Safety Management

- PTA – Pre Task Analysis
- RCI – Root Cause Investigation
- SME – Subject Matter Expert
- SOP – Standard Operating Procedure
- QA/QC – Quality Assurance / Quality Control

References

1. ISO 15189, CFR - Code of Federal Regulations Title 21, International Laboratory Accreditation Cooperation, Good Laboratory Practice Standards; Toxic Substances Control Act (TSCA).
2. ISO 9001 - Quality management, PSM OSHA 3133.
3. Preventative Maintenance – ANSI.
4. OSHA 3071 - 2002 (Revised).
5. ANSI Z358.1-2009 Standard.
6. OSHA 1910.147.
7. OSHA 1910.132.
8. NFPA 70E-2000, www.nfpa.org.
9. OSHA 3132 - 2000.

Chapter 2

Thermodynamics as a Tool for Laboratory and Chemical Safety in the Undergraduate Chemistry Curriculum

Elmer B. Ledesma[*,1] **and Mary K. Moore**[2]

[1]University of St. Thomas, Department of Chemistry and Physics, Houston, Texas 77006
[2]Eastman Chemical Company, Kingsport, Tennessee 37662
*Tel.: +1-713-831-7810; E-mail: ledesme@stthom.edu

Recent laboratory accidents in academic chemistry laboratories have highlighted the importance of the teaching and implementation of laboratory and chemical safety. Traditionally, the undergraduate chemistry curriculum is centered on the teaching of the core subfields of chemistry: analytical chemistry, inorganic chemistry, organic chemistry, and physical chemistry. Laboratory and chemical safety on the other hand are taught in a cursory fashion, usually in the form of a safety video and/or presentation at the beginning of the laboratory course and never revisited as the focus of the course shifts towards mastering the subject matter. Part of the problem is the lack of pedagogical resources that specifically address laboratory and chemical safety. Another is the devaluation of laboratory and chemical safety by many instructors, since they consider teaching fundamental chemical concepts as higher priority. Laboratory and chemical safety, however, can actually be taught in conjunction with the teaching of fundamental concepts. We demonstrate this with thermodynamics, a fundamental concept covered in general and physical chemistry courses. We illustrate, via examples, how laboratory and chemical safety can be used as practical applications of thermodynamics.

Introduction

After investigating a laboratory explosion incident that occurred in the Chemistry and Biochemistry Department at Texas Tech University in January 2010, the Chemical Safety Board (CSB) issued a report that detailed not only deficiencies, key findings, and recommendations relevant to the examinee, but could also be applicable to the entire academic community engaged in scientific and engineering research (1). The CSB primarily investigates accidents at industrial settings, but the Texas Tech incident represented its first investigation of a chemical accident in an academic environment (2). In contrast to the industrial workplace where safety procedures and systems are well entrenched, researchers-professors, postdoctoral associates, technicians, graduate and undergraduate students-working in academic laboratories have had limited to no experience and training in laboratory and chemical safety. This lack of exposure and training has resulted in frequent occurrences of laboratory accidents that, in addition to the Texas Tech incident, include recent chemical accidents that have involved a fatality (2008 fatal burning of a UCLA staff member), injury (2010 hydrogen gas blast at the University of Missouri-Colombia injuring a staff scientist, a graduate student, and two postdocs), and property damage (2010 fire at Southern Illinois University causing significant damage to a chemistry laboratory). In response to the CSB report on Texas Tech, the American Chemical Society's (ACS) Committee of Chemical Safety developed guidelines addressing the gap in safety standards between industrial and academic environments (3). These guidelines are directed toward personnel working in chemical laboratories.

The main foci of principal investigators who manage research programs are obtaining grant funding and publishing research. Postdoctoral associates and graduate students who conduct the research for the principal investigators are under enormous pressure to generate data, and as a consequence design experiments highlighting the experimental needs and may not address the safety aspects involved. Moreover, graduate students who are first starting out to conduct research have limited laboratory and chemical safety training in their undergraduate studies. The principal focus of the undergraduate chemistry curriculum is to teach the fundamental concepts of the core subfields-analytical, inorganic, organic and physical chemistry-of the discipline, and many instructors generally place very little importance to the teaching of safety. In addition, the instructors themselves most likely have not had formal training or taken a course in laboratory and chemical safety. As the next generation of faculty and students arrive with little to no exposure in safety, this cycle will continue, resulting in more laboratory and chemical accidents. In order for the academic research culture to change and thereby lessening the occurrences of safety incidents, students should be exposed to laboratory and chemical safety training throughout their undergraduate chemistry studies to better prepare and equip them upon entering graduate programs and beyond.

Unlike for the chemistry subfields, there is a dearth of pedagogical resources that deals specifically with the teaching of laboratory and chemical safety to undergraduates (4–8), and this is probably a major reason why there are no formal courses on the topic in the undergraduate chemistry curriculum. To circumvent this issue, we propose that undergraduate students can be taught laboratory and chemical safety principles in conjunction with the teaching of the fundamental concepts covered in relevant chemistry courses. More specifically, laboratory and chemical safety can be considered and employed as practical applications of the fundamental concepts that are covered in lecture courses. We demonstrate this approach with thermodynamics, a fundamental concept that is treated in an elementary fashion in freshman general chemistry courses and in more detail in the physical chemistry courses that focus on thermodynamics and kinetics. We have chosen to use thermodynamics as an illustration because it (and reaction kinetics) plays a central role in governing chemical processes, whether they involve physical or chemical changes. By applying the fundamental concepts of thermodynamics to laboratory and chemical safety, undergraduates will not only actively learn the subject matter, but will also appreciate how the abstract concepts in thermodynamics can actually be used to best design and conduct experiments by considering the safety issues involved.

We demonstrate the application of thermodynamic principles to laboratory and chemical safety through three generalized laboratory illustrations: (1) raising the temperature of an organic solvent in a vessel containing air; (2) mixing of acidic (or basic) solutions with water; and (3) combustible vapor mixtures. These three illustrations involve typical procedures that undergraduate chemistry students may encounter in either a laboratory course setting or an undergraduate research setting. For each scenario we discuss the relevant thermodynamic principles involved and the associated laboratory and chemical safety.

Illustrations

Illustration 1: Raising the Temperature of an Organic Solvent in a Vessel Containing Air

An organic solvent is either introduced or stored in a vessel containing air. The system temperature is then raised to a certain value. The solvent and air reach equilibrium with some of the solvent existing as a vapor. Such a scenario can occur during a routine synthetic procedure or can simply be a container containing an organic solvent being left out in the heat or a storage room with faulty environmental controls.

If the composition of the gas mixture (solvent vapor and air) in the vessel is within a specified range, it will ignite and burn if exposed to a flame or a spark. This specified range of composition is determined by the solvent's lower flammability limit (LFL) and upper flammability limit (UFL). If the composition is lower than the LFL or above the UFL, the gas mixture will not burn. The gas mixture is flammable, hence constituting an explosion hazard, only when its composition is between the LFL and UFL. Values of LFL and UFL of compounds can be found in handbooks (*9, 10*). A common unit used to report LFL and UFL is volume percent, which is equivalent to mole percent assuming that the gases obey the ideal gas equation of state, an equation familiar to students taking general chemistry.

To determine the explosion hazard posed by the situation mentioned above, an estimate of the gas mixture composition is needed. This can be achieved through phase equilibrium thermodynamics. Freshman students taking a course in general chemistry learn about the concept of vapor pressure and how to calculate the vapor pressure of a compound at a given temperature using the Clausius-Clapeyron equation,

$$\log_e\left(\frac{P_1^*}{P_2^*}\right) = \frac{\Delta H_{vap}}{R}\left(\frac{1}{T_2} - \frac{1}{T_1}\right) \tag{1}$$

where P_1^* and P_2^* are the vapor pressures at temperatures (in K) T_1 and T_2, respectively, ΔH_{vap} is the enthalpy of vaporization (in J mol^{-1}), and R is the gas constant (in J mol^{-1} K^{-1}). To determine the vapor pressure P_1^* at temperature T_1, the vapor pressure P_2^* at temperature T_2 must be known. The normal boiling point can be used for T_2 because at this temperature P_2^* is equal to atmospheric pressure. The enthalpy of vaporization, ΔH_{vap}, is also needed. Values of ΔH_{vap} can mostly be found in handbooks (9) and online databases (11). If data cannot be found, an approximate value for ΔH_{vap} can be obtained by calculating the change in enthalpy for the phase change process, liquid $\rightarrow$ vapor, a procedure that is covered in the general chemistry course. Standard enthalpies of formation can be used for the enthalpies of the liquid and vapor states and these values are well tabulated for many compounds (9–11).

Once the vapor pressure of the solvent at the specified temperature has been determined, an estimate of the concentration of solvent vapor in the gas mixture can be calculated through the following equation covered in a general chemistry course,

$$y_{solvent} = \frac{P_{solvent}^*}{P_{total}} \tag{2}$$

where $y_{solvent}$ is the mole fraction of solvent vapor in the gas mixture, $P^*_{solvent}$ is the vapor pressure of the solvent at the given temperature, and P_{total} is the equilibrium pressure of the gas mixture assumed to be the sum of the solvent vapor pressure and atmospheric pressure. The mole fraction value is converted to a percentage and then compared to the LFL or UFL in mole percent to determine whether or not the gas mixture constitutes an explosion hazard.

As an example, suppose 200 mL of methanol is introduced into a 3-L flask containing air. The flask is sealed and the system temperature is increased to 45°C. The current air pressure is 1 atm. Assuming that the contents in the flask reach equilibrium, determine whether or not the flask poses a potential explosion hazard. The flammability limits for methanol are 6.0-36 v/v% (10). Methanol has a normal boiling point of 64.6°C and its standard enthalpy of formation for the vapor and liquid states are -201.0 kJ mol^{-1} and -239.2 kJ mol^{-1}, respectively (10). ΔH_{vap} is the difference between the vapor and liquid state standard enthalpies of formation:

$$\Delta H_{vap} = (-201.0) - (-239.2) = 38.2 \; kJ \; mol^{-1} = 38000 \; J \; mol^{-1}$$

Using Equation (1) with $R = 8.314 \; J \; mol^{-1} \; K^{-1}$, $T_2 = 64.6°C = 337.75 \; K$ and $P_2^* = 1$ atm (normal boiling point of methanol), the vapor pressure, P_1^*, at $T_1 = 45°C = 318.15 \; K$ is determined:

$$\log_e\left(\frac{P_1^*}{1}\right) = \frac{38000}{8.314}\left(\frac{1}{337.75} - \frac{1}{318.15}\right)$$

$$\log_e P_1^* = -0.8337$$

$$\therefore P_1^* = e^{-0.8337} = 0.434 \ atm$$

The total pressure exerted by the contents of the flask is the sum of the vapor pressure of methanol and the atmospheric pressure (air was initially present in the flask). Therefore, using Equation (2), the mole fraction of methanol in the gas mixture can be calculated:

$$y_{methanol} = \frac{0.434}{0.434 + 1} = 0.303 \ or \ 30.3 \ mol\%$$

The calculated value of 30.3 mol% is within the flammability limits of methanol. Therefore, for the situation described in this example, the flask is a potential explosion hazard.

Illustration 2: Mixing of Acidic (or Basic Solutions) with Water

The current room temperature in a laboratory is $T_{room}°$C. A student is to perform a dilution by mixing $V1$ mL of pure nitric acid, HNO_3, with distilled water to give a final solution containing x wt% of HNO_3. In order to assess a potential safety hazard associated with a rise in temperature upon mixing, the student needs to determine what maximum temperature, $T°$C, can be attained.

General chemistry students are taught that when mixing solutions of a concentrated acid and water together, the acid must be added to water. If the reverse procedure were to take place, a straightforward laboratory demonstration would clearly show the drastic rise in temperature as the mixed solution boils. In upper division physical chemistry classes, undergraduate students learn in detail the fundamental concept of enthalpy of mixing, which is the basis for the acid-water mixing scenario. Most students in physical chemistry do not cover the practical aspects of the enthalpy of mixing, as instructors often focus on abstract and theoretical aspects. However, the scenario depicted above affords an excellent practical application of the change in enthalpy of mixing.

In order to determine the final temperature attained upon mixing, the composition of the product solution needs to be determined. An acid solution of known concentration is to be diluted with water to give a product solution of a specified concentration. The basis behind this determination is the law of conservation of mass, a fundamental concept first encountered in freshman general chemistry courses. The mass of HNO_3 in the acid solution before mixing is equal to the mass of HNO_3 in the final product solution after mixing. As a result, a simple mass balance equation on HNO_3 can be set up. In the laboratory, volumes of solutions are much easier to deal with than their masses due to the preponderance of volumetric glassware. To carry out the mass balance calculation, volumes need to be converted to masses and the densities of the solutions are therefore needed. Densities (and specific gravities) of pure substances and aqueous solutions are well documented (9–12). Once the density (in g mL^{-1}), ρ_1, of the pure HNO_3 solution at temperature T_{room}°C has been obtained, its mass (in g) m_1 can be calculated from its volume V_1 in mL:

$$m_1 = \rho_1 V_1 \tag{3}$$

The mass of HNO_3 in this solution is equal to the mass of HNO_3 in the final solution once the mixing has taken place:

$$m_1 = (m_1 + m_{water})\left(\frac{x}{100}\right) \tag{4}$$

The unknown in the equation above is m_{water}, the mass (in g) of water needed to dilute the original acid solution to give a product solution with the desired concentration. From the mass of water, the volume of water can be calculated.

Now that the composition of the product solution has been calculated, the maximum temperature attained after the mixing can be determined using an energy balance. This is accomplished through the First Law of Thermodynamics, a concept that is covered in an elementary fashion in general chemistry and treated in more detail in physical chemistry. The system described here is at constant pressure and assumed to be adiabatic. As a result, the First Law for this system can be written as,

$$\Delta H = 0 \tag{5}$$

where ΔH is the change in enthalpy (in J) between the final product solution after the mixing and the water and original acid solution before the mixing plus the enthalpy of mixing. ΔH can therefore be expressed as

$$\Delta H = \Delta H_{mix} + \Delta H_{solution} \tag{6}$$

where ΔH_{mix} is the enthalpy of mixing (in J) and $\Delta H_{solution}$ is the change in enthalpy (in J) between the final product solution after mixing and the water and the original acid solution before the mixing. $\Delta H_{solution}$ can be written in terms of temperature, the total mass of the final solution, and the specific heat capacity, $\hat{c}$, of the final solution:

$$\Delta H_{solution} = m_{solution} \int_{T_{room}}^{T} \hat{c}\, dT \tag{7}$$

To solve this problem ΔH_{mix} is needed. These values for common acids at different dilutions in water can be found in handbooks (*10*). Once found in a handbook, Equation (6) can then be solved to determine the final temperature, T, which can then be used to assess any potential hazards and to identify what, if any, safety precautions need to be considered before performing the dilution.

As an example, a student needs to make a 50 wt% solution of nitric acid by diluting 500 mL of pure nitric acid with distilled water. The temperature in the laboratory is 20°C. In order to assess potential safety hazards, the student is to calculate the final temperature of the resulting solution that can be attained. The amount of water needed for the dilution is first found through a mass balance on HNO_3. The mass of HNO_3 in the final solution is equal to the mass of HNO_3 in the pure nitric acid solution. The density of pure nitric acid at 20°C is 1.5129 g mL^{-1}

(12). Using this density and the volume of pure nitric acid, the mass of HNO_3 in the final solution can be calculated:

$$m_1 = 1.5129(500) = 756.45 \ g$$

Using Equation (4) the mass of water needed to dilute the 500 mL pure nitric acid solution to produce a 50 wt% solution of HNO_3 is determined as follows:

$$756.45 = (756.45 + m_{water})\left(\frac{50}{100}\right)$$

$$\therefore \ m_{water} = 756.45 \ g$$

Since the density of water at 20°C is 0.99821 g mL^{-1} (10), therefore the volume of water needed to perform the dilution is 758 mL. Now that the mass balance has been determined, the energy balance, Equation (6), needs to be solved for the final temperature. To calculate ΔH_{mix} the molality of the final solution needs to be determined. Molality is covered in freshman chemistry courses and it is defined as follows:

$$molality = \frac{moles \ of \ solute}{kg \ of \ solvent} \tag{8}$$

Using this equation, the molality of HNO_3 in the final solution is easily obtained:

$$molality = \frac{756.45/63.02}{0.75645} = 15.9$$

The CRC Handbook of Chemistry and Physics (10) lists enthalpy of mixing for a number of common acids at different molalities. Using the molality value obtained for this problem, an interpolation of the CRC data gives $\Delta H_{mix} = -26 \ kj \ per \ mole \ of \ solute$. The number of moles of solute is $756.45/63.02 = 12 \ moles$, therefore $\Delta H_{mix} = -312 \ kJ$. The specific heat capacity of 50 wt% HNO_3 aqueous solution is $2.72 \ J \ g^{-1} \ °C^{-1}$ (12). Using Equation (7) in Equation (6), the energy balance is solved as follows:

$$0 = -312000 + (756.45 + 756.45)\int_{20}^{T} 2.72 \ dT$$

$$312000 = (1512.9)(2.72)(T - 20)$$

$$\therefore T = 95.8°C$$

Therefore, diluting 500 mL of a pure nitric acid solution with 758 mL of distilled water will result in a solution whose temperature will reach 95.8°C.

Illustration 3: Combustible Vapor Mixtures

A reaction vessel (or container) is charged with a stoichiometric gas mixture of an organic compound and air. The initial temperature and pressure are 298 K and 1 atm. The mixture is then ignited with a spark or a flame. Such a scenario is typical for performing combustion experiments in an advanced physical chemistry laboratory course or in graduate research. The major safety hazard when conducting such an experiment is the potential compromising of the structural integrity of the reaction vessel due to the rapid rise in temperature and pressure of the contents in the vessel as the combustion reaction proceeds. Before conducting the experiment, it is necessary to determine the maximum possible temperature and pressure that can be obtained. Knowing these two values will determine what sort of material will be used for the reaction vessel.

The maximum possible temperature that can be obtained in a combustion reaction is called the adiabatic flame temperature, T_{ad}. Once this temperature and the total amount of products produced are calculated, the maximum pressure possible can be determined using the ideal gas equation of state. The total amount of products produced is calculated through reaction stoichiometry, a fundamental chemistry concept covered in general chemistry. To determine T_{ad}, the First Law of Thermodynamics is applied. For the constant-volume system discussed in this illustration and assuming adiabatic conditions, the First Law is expressed as,

$$\Delta U = U_{products} - U_{reactants} = 0 \tag{9}$$

where ΔU is the change in internal energy (in J) between the products, $U_{products}$, and reactants, $U_{reactants}$. In a general chemistry course and treated in more detail in physical chemistry lecture courses, students learn the relationship between internal energy and enthalpy,

$$H = U + PV \tag{10}$$

where H is enthalpy, P is pressure, and V is volume. Using this relation, the First Law can be expressed in terms of enthalpy, pressure, and volume as follows:

$$H_{products} - P_{products}V - H_{reactants} + P_{reactants}V = 0 \tag{11}$$

Here $H_{reactants}$ and $H_{products}$ are the enthalpies (in J) of the reactants and products, respectively, $P_{reactants}$ and $P_{products}$ are the total pressures (in Pa) of the reactants and products, respectively, and V is the reactor volume (in m^3). Using the ideal gas equation of state to eliminate the PV terms and rearranging, the First Law is thus written as,

$$H_{reactants} - H_{products} - R\left(N_{reactants}T_{reactants} - N_{products}T_{ad}\right) = 0 \tag{12}$$

where R is the gas constant (in J mol^{-1} K^{-1}), $T_{reactants}$ is the initial temperature of the reactant mixture (in K), and $N_{reactants}$ and $N_{products}$ are the total number of moles of reactants and products, respectively. To determine T_{ad} the enthalpy of the reactants, $H_{reactants}$, and the enthalpy of the products, $H_{products}$, need to be determined. This is accomplished through the following equation,

$$H = \sum_{i=1}^{n} N_i \underline{H}_i \tag{13}$$

where N_i is the number of moles of species i and $\underline{H}_i$ is the molar enthalpy (in J mol^{-1}) of species i and is calculated from the following equation:

$$\underline{H} = \Delta \underline{H}^\circ + \int_{298}^{T} C_P \, dT \tag{14}$$

In this equation $\Delta \underline{H}^{\circ}$ is the standard enthalpy of formation (in J mol^{-1}), T (in K) is the temperature of the reactants or products (T_{ad}), and C_P is the molar heat capacity at constant pressure (in J mol^{-1} K^{-1}).

The adiabatic temperature can thus be calculated once all the necessary pieces of information have been gathered. The molar heat capacity is usually a function of temperature, and the calculation to determine the adiabatic flame temperature will require a numerical procedure to determine its value. Once T_{ad} has been calculated, the maximum pressure that can be achieved in the reaction vessel is then calculated via the ideal gas equation of state.

As an example, a 1-L reaction vessel is charged with a stoichiometric mixture of methane and air at 298 K and 1 atm. If the mixture is ignited, determine the maximum temperature and pressure possible. To calculate the maximum temperature possible, Equation (12) needs to be solved for T_{ad}. Since air can be approximated as composed of 21 mol% O_2 and 79 mol% N_2 (*10*), the balanced chemical equation for the complete combustion of methane in air is the following,

$$CH_4 + 2O_2 + 7.52N_2 \rightarrow CO_2 + 2H_2O + 7.52N_2$$

where the total number of moles of reactants and products are equal to one another, i.e. $N_{reactants} = N_{products} = 10.52 \, moles$. As such, the third term on the left hand side in Equation (12) can be written as follows:

$$R\left(N_{reactants}T_{reactants} - N_{products}T_{ad}\right) = 8.314(10.52)(298 - T_{ad})$$

At the initial conditions of 298 K and 1 atm, the standard enthalpies of formation of the reactants are -74.6 kJ mol^{-1} for methane and 0 kJ mol^{-1} for oxygen and nitrogen (*10*). Since the initial temperature of the reactants is 298 K, the integral in Equation (14) is zero for methane, oxygen, and nitrogen. Therefore, $H_{reactants}$ as computed through Equation (13) is the following:

$$H_{reactants} = (1)(-74600) + (2)(0) + (7.52)(0) = -74600 \, J$$

The constant pressure heat capacities (in J mol^{-1} K^{-1}) of the products depend on temperature and are as follows (*13*):

$$C_P^{CO_2} = 22.243 + (5.977 \times 10^{-2})T + (-3.499 \times 10^{-5})T^2 + (7.464 \times 10^{-9})T^3$$

$$C_P^{H_2O} = 32.218 + (0.192 \times 10^{-2})T + (1.055 \times 10^{-5})T^2 + (-3.593 \times 10^{-9})T^3$$

$$C_P^{N_2} = 28.883 + (-0.157 \times 10^{-2})T + (0.808 \times 10^{-5})T^2 + (-2.871 \times 10^{-9})T^3$$

Using the standard enthalpies of formation of the products (*10*) and the expressions for their constant pressure heat capacities given above, the total enthalpy of the products can be written as follows using Equation (12):

$$H_{products} = (1)H_{CO_2} + (2)H_{H_2O} + (7.52)H_{N_2}$$

where Equation (14) is used to determine the individual product enthalpies:

$$H_{CO_2} = -393500 + \int_{298}^{T_{ad}} C_P^{CO_2} \, dT$$

$$H_{H_2O} = -241800 + \int_{298}^{T_{ad}} C_P^{H_2O} \, dT$$

$$H_{N_2} = 0 + \int_{298}^{T_{ad}} C_P^{N_2} \, dT$$

Equation (12) can now be written as follows:

$$[-74600] - \left[(1)\left(-393500 + \int_{298}^{T_{ad}} C_P^{CO_2} \, dT \right) + (2)\left(-241800 + \int_{298}^{T_{ad}} C_P^{H_2O} \, dT \right) \right.$$

$$\left. + (7.52)\left(\int_{298}^{T_{ad}} C_P^{N_2} \, dT \right) \right] - 8.314(10.52)(298 - T_{ad}) = 0$$

This equation is a nonlinear equation and a numerical method needs to be used in order to solve for the unknown T_{ad}. A FORTRAN 77 routine is used to determine the zero of the nonlinear equation given above (*14*). Solving for the unknown yields a value of T_{ad} = 2986 K.

Calculating the maximum pressure possible is now easily determined. Using the ideal gas equation of state and noting that $N_{reactants} = N_{products}$, the final pressure is calculated as follows:

$$P_{final} = P_{initial} \left(\frac{T_{ad}}{T_{initial}} \right)$$

$$P_{final} = (1) \left(\frac{2986}{298} \right)$$

$$\therefore P_{final} = 10.0 \; atm$$

The calculations therefore show that during combustion of the mixture, the maximum temperature and pressure attained are far greater than the initial temperature and pressure.

Conclusions

Through three illustrations involving laboratory techniques and procedures that are encountered in the undergraduate chemistry curriculum or in research, we have demonstrated how laboratory and chemical safety can be used as practical applications of the fundamental concepts covered in thermodynamics. The fundamental concepts of thermodynamics are generally treated in a theoretical and abstract manner in general and physical chemistry courses wherein students are required to determine molecular or reaction properties. In most instances, the treatment of thermodynamics in the undergraduate chemistry curriculum does not illustrate its practical relevance to laboratory and chemical safety. By applying the fundamental concepts of thermodynamics to laboratory and chemical safety, undergraduates will not only actively learn the subject matter, but will also appreciate how the abstract concepts in thermodynamics can actually be used as a tool for safety and as a guide when performing experiments either in an undergraduate laboratory course or in research.

References

1. U.S. Chemical Safety and Hazard Investigation Board. *Texas Tech University: Laboratory Explosion*; Final Report, No. 2010-05-I-TX; Washington, DC, 2010.
2. Johnson, J. School Labs go under Microscope. *Chem. Eng. News* **2010**, *88*, 25–26.
3. American Chemical Society. *Identifying and Evaluating Hazards in Research Laboratories*; Washington, DC, 2013.
4. American Chemical Society. *Safety in Academic Chemistry Laboratories*, 7th ed.; Washington, DC, 2003; Vol. 1.
5. Furr, A. K. *CRC Handbook of Laboratory Safety*; 5th ed.; CRC Press: Boca Raton, FL, 2000.
6. National Research Council. *Prudent Practices in the Laboratory: Handling and Management of Chemical Hazards*, Updated Version; National Academies Press: Washington, DC, 2011.
7. Hill, R. H.; Finster, D. C. *Laboratory Safety for Chemistry Students*; John Wiley & Sons, Inc.; Hoboken, NJ, 2010.
8. Crowl, D. A.; Louvar, J. F. *Chemical Process Safety: Fundamentals with Applications*; Prentice Hall: Englewood Cliffs, NJ, 1990.
9. Speight, J. G. *Lange's Handbook of Chemistry*, 16th ed.; McGraw-Hill: New York, NY, 2005.
10. Lide, D. R. *CRC Handbook of Chemistry and Physics*, 86th ed.; CRC Press: Boca Raton, FL, 2005.
11. http://webbook.nist.gov/.
12. Perry, R. H.; Green, D. W. *Perry's Chemical Engineers' Handbook*, 7th ed.; McGraw-Hill, New York, NY, 1997.
13. Sandler, S. I. *Chemical and Engineering Thermodynamics*, 4th ed.; John Wiley & Sons: New York, NY, 2006.
14. Shampine, L. F.; Watts, H. A. *FZERO, A Root-Solving Code*; Report SC-TM-70-631, Sandia Laboratories: September 1970.

Chapter 3

Considerations for Scale-Up – Moving from the Bench to the Pilot Plant to Full Production

Frankie Wood-Black*

Principal – Sophic Pursuits, Inc., 6855 Lake Road, Ponca City, Oklahoma 74604
***E-mail: fwblack@sophicpursuits.com**

When developing technologies, there are a number of steps required between the initial concept and completion of the final production plant. These steps include the development of the commercial process, optimization of the process, scale-up from the bench to a pilot plant, and from the pilot plant to the full scale process. While the ultimate goal is to go directly from process optimization to full scale plant, the pilot plant is generally a necessary step. Reasons for this critical step include: understanding the potential waste streams, examination of macro-processes, process interactions, process variations, process controls, development of standard operating procedures, etc. The information developed at the pilot plant scale allows for a better understanding of the overall process including side processes. Therefore, this step helps to build the information base so that the technology can be permitted and safely implemented. This paper focuses on the specific needs of the operating plant to allow a new technology to be implemented.

Introduction

Recently, there was an internet survey; asking individuals to describe the difference between a scientist and an engineer (Anne Marie Helmenstine, n.d.). One responder put it this way: "The difference is illustrated in the relationship between Mr. Spock and Mr. Scott in *Star Trek*. Spock's interest in a problem ends when he has determined that a solution is feasible, Scotty's ends when he has produced a solution that works." For Mr. Spock, proof of concept is all that is needed. However, for Mr. Scott there are a number of challenges that need to be overcome in order to put the solution into practice. The operating plant is filled with Mr. Scotts. Recall, Mr. Scott's frustrations with Mr. Spock, when Mr. Spock did not seem to take into account all of the parameters, boundaries, and constraints that he was forced to deal with in order to provide a working solution. These constraints are exactly what the operating plant is going to want to understand prior to beginning any full scale operation. The operating plant is worried about the technical and practical aspects of making the final product.

For researchers and application engineers, the transition between proof of concept and application is not as simple as that portrayed in an episode of *Star Trek*. The engineer or plant operator can see some of the challenges of concept implementation but not all. There may be additional economic or regulatory constraints. Each of the constraints may require a specific solution or present an additional challenge. Each potential solution, in turn, may present additional issues which will require further thought or applications of other ideas. These are just the obvious hurdles, i.e. ones that can be reasonably anticipated by a project team. However, there generally some of the "hidden" or not readily apparent hurdles may ultimately scrap a conceptual solution because it is no longer economically feasible. These "hidden" hurdles may only become apparent as the process moves from the bench scale to a larger scale.

In addition to the engineering considerations of process conditions, and materials management; there are environmental, safety and regulatory hurdles that must be addressed. A bench scale reaction may only produce grams of the potential product. More material may be required for toxicity, environmental fate, and chemical testing. The specific process conditions for the economic viable process have to be determined, e.g. pressure, temperature, time, etc. Additionally, an evaluation of the process intermediates and potential contaminants in the supply change and/or products has to be completed. These side reactions, contaminates and unintended processes may have a dramatic impact on the results of the required testing, ultimate use, and viability of the process. For example: if the ultimate product is a drug formulation – the contaminates may impact the efficacy of the drug and therefore require additional process steps to remove the contaminate or a change in the source of the raw material.

Hence, the need for an applied research phase between the pure concept and implementation. This applied research phase is usually referred to as the scale-up, moving from a bench scale to full production scale. For many, the applied research phase is conducted at a scale somewhere in between, the pilot plant scale.

The End Game

For the ultimate production scale plant, there are a number of questions that need to be addressed even before the final overall concept is presented to the decision makers, i.e. the individuals that are going to provide the funding or giving approval to the project. The decision makers include the investors, the construction contractors, the regulators, and the general public potentially impacted by the

Figure 1. Titration of Cr(VI) with Fe(II) to the phenanthroline endpoint under strongly acidic conditions. Phenanthroline is blue-green in the presence of Fe(III) and Cr(VI), but a red chelate forms from excess Fe(II) as when the Cr(VI) is fully reduced.

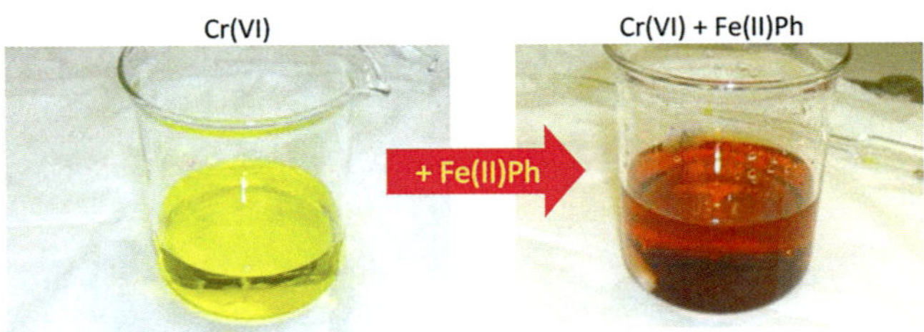

Figure 3. The red Fe(II) phenanthrolene complex remains red when added to a neutral Cr(VI) solution, indicating that Cr(VI) is not neutralized by chelated Fe(II).

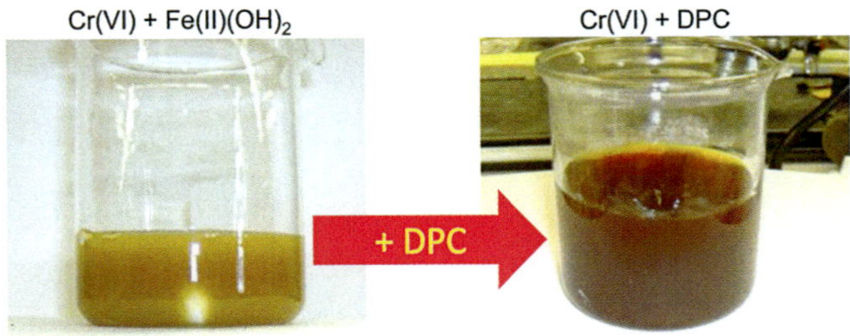

Figure 4. Fe(II) under alkaline conditions forms a yellow-brown precipitate, probably Fe(OH)$_2$. Adding Cr(VI), then diphenyl carbazide (DPC) produces a dark cloudy red color, indicating that much of the Cr(VI) was not neutralized.

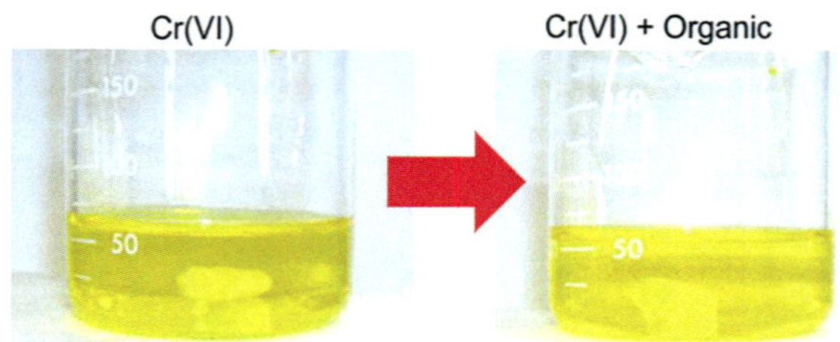

Figure 5. Organic compounds, including isopropanol, acetone, glycolic acid, oxalic acid or acetic acid, were added to chromate solutions at near neutral pH. No reduction of Cr(VI) occurred within an hour.

Figure 6. Reduction of Cr(VI) by sodium bisulfite solution.

Figure 7. Reduction of Cr(VI) by potassium iodide.

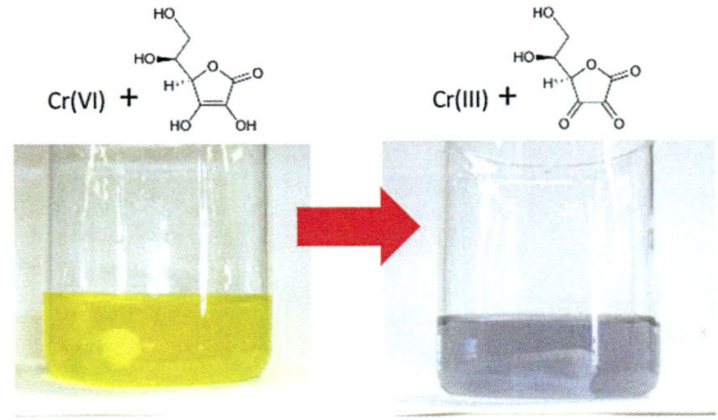

Figure 8. Reduction of chromate by ascorbic acid.

Figure 9. Cr(VI)/silica catalyst is reduced by ascorbate to the green color of Cr(III)/silica, but if titanium is present it then chelates with the oxidized ascorbate to form a red complex.

Figure 10. Diphenyl carbazide indicator turns a deep purple in the presence of Cr(VI). It can be used to detect catalyst dust on surfaces.

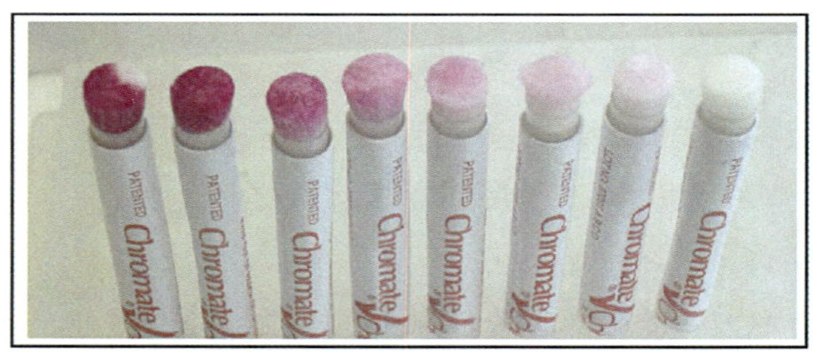

Figure 11. ChromateCheck calibrated test swabs to detect Cr(VI) on a surface. Note: 3 mcg Cr(VI) per swipe would require action.

project. The concerns of the decision makers are varied. The questions that may be posed prior to project approval include but are not limited to:

- What are the construction costs?
- What are the continuing operating costs?
- What are the impurities that may be found in the product?
- What by-products are produced by this process?
- What are the operating conditions?
- What are the costs of the raw materials?
- How available are the raw materials?
- Where is the location of the plant in relation to the distribution channels and/or raw materials?
- How energy intensive is the process?
- What are the waste products?
- What are the environmental considerations – emissions, wastes, potential exposures, etc.?
- How efficient is the process?
- What is the competitive technology?
- Are the raw materials, by-products, wastes, etc. regulated?
- How safe is the process?
- What are the hazards associated with the process?
- What are the hazards associated with the raw materials?
- What are the hazards associated with the shipping, storage, and handling of the raw materials, wastes and final products?
- What controls are going to be necessary?
- What is the overall configuration or footprint of the process?

The list of questions can continue. The answers to these questions need to be understood and articulated in business proposals, permit applications, community outreach sessions, and other documents in order for the initial construction of the new facility or modification of an existing facility can even begin. Yet, while some of the questions may be answered using bench scale data or in the general process description; many of the answers can only be determined during the pilot plant stage of the technology development. And, there may be some things that are still unknown, i.e. what is it about the proposed process that could be a potential project hurdle?

Depending on the specific questions to be addressed, one can find a variety of case studies in the literature focusing on the various individual aspects of the pilot plant process. For example: the Royal Society of Chemistry published a book on "Industrial Chemistry Case Studies: Industrial Processes in the 1990's." In this book several aspects of moving from concept to production are evaluated with examples from the pharmaceutical, steel, and polymer industries.

In 2013, H. A. Aziz, et.al. published a paper in the Proceedings of the 6th International Conference on Process Systems Engineering focusing on the overall management of chemical information from the pilot plant. In this case study, the goal was to obtain, store, and process specific information gathered during the operation of the pilot plant to develop the required health and safety information needed to properly implement the process safety management standard in the full production plant.

Economic case studies abound. These studies focus on optimizing the overall industrial process, supply chain, and energy use of the final design. There are several companies and engineering firms whose entire focus is on the evaluation of the economics associated with the final process design.

As the emphasis here is to gain insight into the overall process of moving from bench to the industrial process, it is impossible to focus on a single aspect in detail. What is important to note is that scale-up requires a team approach. The scale-up team will need to include members from a variety of disciplines. The team needs to include: knowledge experts from the bench; end users of the information gained from the pilot plant studies; quality experts; finance; health, environment and safety; information technology; suppliers, etc. The team will need to evaluate the specific questions to be addressed by the scale-up process.

The Typical Process Diagram

Conceptually, a process can be viewed as a collection of inputs, i.e. the reactants, catalysts, solvents, which are reacted to form a new primary product with potential by-products and recoverable materials. In a perfect world, the reaction goes as planned and there are only the desired outputs. Figure 1 provides a simplistic view of this conceptual process. Yet, this process does not account for reality.

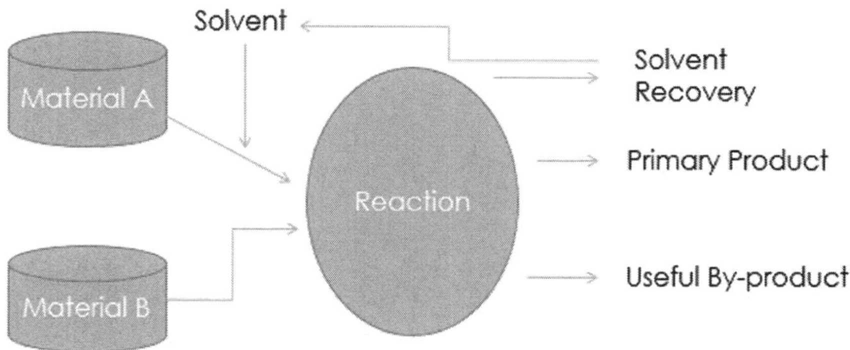

Figure 1. Typical conceptual process diagram.

The conceptual process is missing a number of key considerations (Figure 2). Processing is not perfect, thus, there are emissions and waste materials. Theoretical or predicted yields are not achieved. There are unanticipated wastes. There are processing agents that may be required to protect the process equipment from corrosion or to aid in overall reactions, e.g. emulsifiers or emulsion breakers.

The energy or utility inputs are not considered. The side processes of purification of products, by-products, and solvent recovery are not included. These may or may not be all of the additional process considerations that need to be addressed. The conceptual process needs to be transformed into a real process where these considerations are evaluated and understood.

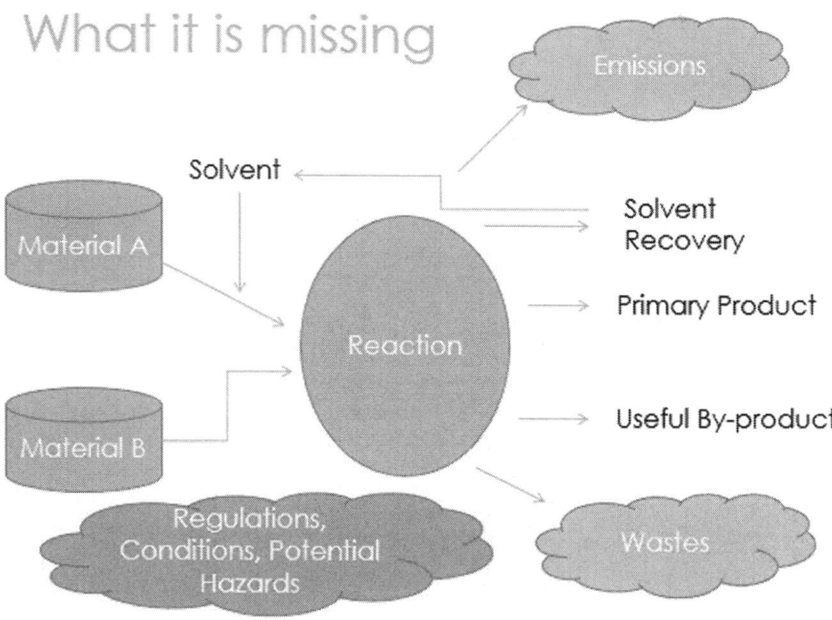

Figure 2. Examples of what may be missing from the conceptual process design.

The pilot plant stage used to help identify the processing needs, the additional requirements, and identification of the unknowns. For example: a bench scale reaction may result in an impurity that can be carried with the primary product. Because of the scale of the process this small impurity may not be a major concern. On a production scale this impurity may impact the overall product to the point it cannot be used for the intended purpose, e.g. a heavy metal contaminate in a drug precursor, or a compound that reduces the efficacy of a pharmaceutical. The impurity may have long lasting ecological effects, e.g. an impurity in an agrichemical which would prevent its application on crops intended for human consumption.

The impurity may result in a waste stream that needs to be managed. The characteristics of this waste stream need to be determined and the processes for managing the material addressed. The pilot plant scale will help in this determination as the waste stream will magnified at the larger scale. Caution is still required as even at the pilot plant scale; the magnitude of the unknown may not be large enough for full consideration. This may result in an unintended issue or concern at larger scales.

For example, in the petroleum industry many reactions and processes generate a wastewater stream. Analyses of these wastewater streams at the bench scale may not identify any particular contaminate of concern. Yet, as the process is scaled-up, the concentration of contaminates may now reach regulatory levels. Selenium is one such contaminate that may not be at a high enough concentration to be observed as a contaminate of concern at the bench or pilot plant but as the regulatory limits have been modified over time has become a major concern for many petroleum refineries at the production level.

From the Bench to the Pilot Plant

From the conceptual process flow and process diagrams, there are a number of aspects of the larger process that may be reasonably predicted. For example: initial by-products, solvent needs, potential side processes, e.g. solvent or catalyst recovery, may be identified. These "knowns" will help to design the pilot plant. General processing conditions and reaction chemistries are known. Some potential hazards have been identified. Initial optimization of the process can begin.

The first step in the scale-up process is the design of the pilot plant. From the information obtained from the bench scale process, the design engineer can begin to shape the pilot plant process and scope an overall design. The initial question of this design is "will the conditions at the bench translate to the scale-up?" Thus, the temperature, pressure, mixing, timing of the reaction, controls, etc. all have to be evaluated. Even the materials of the reaction vessel need to be considered. For example: a bench scale reaction may be conducted in a glass vessel. Will the reaction be impacted by conducting the reaction in a steel vessel? Changes in metallurgy or vessel requirements may impact the economics of the scale-up and the ultimate economic feasibility of the overall project.

There are mechanical issues as well to be addressed. If the process requires mixing, or agitation; how is that to be accomplished at the larger scale? Does the shape of the reaction vessel matter? Are mixers going to be required? How is complete mixing going to be evaluated? What happens if complete mixing is not achieved?

The physical state of the reactants, solvents, catalysts, etc. and the material handling properties need to be evaluated. At the bench, the physical state of the material may not be a major concern, yet as the process is scaled-up, how the reactants reach the reaction vessel may require additional processing or specific design considerations. For example: in order to achieve economic feasibility of the process the raw material may need to be purchased in a solid state. It is delivered in flow bins. Yet, the process requires introduction of the reactant in its liquid form. How is the processing facility going to receive the flow bins, and transform the material to the needed form? What are the storage considerations associated with physical state of the raw material? Are there specific storage conditions that need to be met such as temperature or container type?

Finally, the critical parameters of the overall process need to be established and understood. For example, what happens if the temperature is not controlled? What happens if the pressure is not maintained? Is there a pathway to an undesired reaction or outcome? What controls or monitoring systems need to be evaluated? Essentially, the scale-up engineer needs to look at what are the places where things can go wrong. What are the uh-ohs? There is a need to think in the world of, what ifs? The potential hazards and means of hazard mitigation need to be understood and reviewed. With these questions in mind, the designer of the pilot plant begins the overall process of scale-up.

Steps to Scale-Up

As discussed above the end game, i.e. the final production facility, has a large impact on the design and goals of the pilot plant process. The pilot plant development is two-fold, obtaining the necessary information for the final project design and understanding of the unknowns of the process. So far, the general types of questions that may need to be addressed have been discussed, but understanding that these questions exist do not necessarily help the pilot plant designer develop the project goals for the pilot plant nor address how to design a specific plant. Additionally, one pilot plant configuration may not address all the necessary questions.

There are a variety of approaches to overall design. However, from the final operating plant perspective there are three key questions that need to be answered:

1. What is the general process?
2. What are the regulatory requirements?
3. How is it to be controlled and operated?

Answering these three questions, help guide the overall design. Hence, the design steps follow this approach:

- Outlining the general process
- Inclusion of safety and environmental controls
- Development of the processing guidelines and operating system

Using this approach, the goals and the design of the pilot plant may be evaluated. The questions posed earlier can start to be addressed in a more systematic fashion.

Outlining the General Process

From the bench scale, one has a general understanding of the overall reaction. Yet, to reach the processing stage; several decisions must be made and these decisions may be impacted by the anticipated final products and uses. The first decision at least for a chemical process is the determination if the process is batch versus continuous. The overall processing, mechanical considerations, and transport of materials from one step to another and how the process is integrated with the support processes is highly dependent on whether the process is batch or continuous.

This is also the stage were raw materials, metallurgy, piping, transport, auxiliary equipment needs, storage, etc. need to be evaluated and preliminary decisions made. As seen above, the materials used for a bench scale are not necessarily appropriate for the larger scale. Engineering challenges associated with material handling may have to be considered. Finally, the economics of the decisions begin to shape the overall final process.

Inclusion of Safety and Environmental Controls

Once, the general process and engineering decisions are scoped. The safety and environmental consequences of those choices need to be evaluated. Additional, processes may have to be included or changes to the configurations or metallurgy may need to occur. Monitoring and control systems may have to be added. Even the raw material choices may have to be changed.

The general process phase and this phase may occur concurrently or as a type of iterative process. The overall process will be impacted by the needs of the regulatory requirements. There may be specific materials that can or cannot be utilized based on controls associated with the intended use of the product. There may be restrictions on the materials. There may be limitations associated with equipment or processing selection. All of these are essential to the overall process and pilot plant design.

Development of the Processing Guidelines and Operating System

Just as important in the engineering of the physical side of the process, is the engineering of the human side. The process may have a number of integrated controls and computer monitoring, but ultimately there are operators, engineers and technicians that must watch over the process. How the systems are built for this human integration is key to the efficiency and safety of the final operating plant.

Conclusions

As seen there are numerous questions that must be answered between the bench proof of concept and the ultimate implementation of the operating process at full scale production. One utopic view of the scale-up process is a short cut the time and expense by skipping the pilot plant stage and moving directly into the processing facility. Yet, there are several concerns with this view.

The psychoanalytic philosopher Slavoj Zizek stated it very clearly, one of the main dangers associated with people is our refusal to recognize or acknowledge what we know – the unknown known. In the case of process plant design, the unknown known is very much alive and well.

The bench researcher knows that there are impurities that on the plant scale will be wastes, what is unknown is the amount, characteristics, and how this waste will have to be handled. The bench researcher knows that if mixing does not occur properly several unintended outcomes may occur. Yet, on the plant scale, how can this mixing occur properly, how can this be assured, how can we control the unintended outcomes if mixing fails at a particular point in the reaction? The bench research knows that the selection of material A is not optimum on a large scale, yet what are unintended reactions associated with using material B the more economically feasible alternative.

The pilot plant is an essential tool in providing and/or highlighting some of these unknown knowns. It allows for optimization, experimentation related to the "what ifs" on a scale that mitigates some of the risks. It allows for experimentation with metallurgy, control systems, alternatives, etc. The pilot plant provides essential data for construction, permitting, understanding, and operation of the plant scale process. Operation provides for a means of evaluating the human/control interface. Hence, short cutting this process may have severe unintended consequences.

Chapter 4

Management of Change: Pilot Plant Scale

Neal Langerman[*]

**Advanced Chemical Safety, PO Box 152329, San Diego,
California 92195-2329**
[*]E-mail: neal@chemical-safety.com; Tel.: 619-990-4908

Process Safety Management (PSM) is a proven tool for improving safety at units for which OSHA regulations require its application. Experience has shown that PSM improves safety management at all units, whether or not OSHA is applicable. Within PSM is the very powerful tool "Management of Change" (MOC). A research pilot plant and a research laboratory are, by definition units subject to frequent change in operating conditions. These include reactants, pressure, temperature and flow, among others. MOC can be usefully applied to such changing conditions with a reasonable set of guidelines for invoking a MOC safety review. Guidelines for MOC within a pilot plant setting will be described with some brief reviews of cases in which it was applied.

Chemistry research, at the laboratory and pilot plant levels involves a continual evolution of conditions in an effort to achieve the scientific objectives. This necessary series of changes is quite different from change in a production plant, which normally only occurs in discrete and occasional steps. The U.S. Occupational Safety and Health Administration issued the Process Safety Management (PSM) (*1*) rule to address the overall operational safety of large chemical units. While the rule generally does not apply to pilot plant or lab scale operations, its use in those settings can provide a coherent, easily implemented guide for operational safety. A key part of PSM is the "management of change" (MOC). This chapter will examine the application of the OSHA concept of MOC as applied to pilot plants and laboratories (*2*). In particular, the issue of "when is a change really a change" will be examined in detail and a useful control tool will be presented.

The OSHA PSM rule is published in 29 CFR 1910.119. It consists of the following major sections:

- Employees involvement in PSM
- Process safety information
- Process hazard analysis
- Operating Procedures
- Training
- Contractors
- Pre-start-up safety review
- Mechanical integrity
- Hot work permit program
- Management of change
- Incident investigation
- Emergency planning & response
- Compliance audits
- Trade secrets

Langerman (*1*) discusses the application of this Standard to laboratory situations. In 2008, MOC in pilot plants was discussed, but lacked an implementation tool. PSM contains a list of chemicals with trigger threshold values. The triggers are much larger than the quantities of chemicals found in laboratories and most pilot plants, and the list itself is limited. But the protocols contained in PSM and in the MOC paragraph in particular apply to all research situations, whether or not a regulatory trigger is imposed.

Two key definitions are needed to understand the application of MOC and use of the tool to be presented.

- **HIGHLY HAZARDOUS CHEMICAL** means a substance possessing toxic, reactive, flammable, or explosive properties.
- **PROCESS** means any activity involving a highly hazardous chemical including any use, storage, manufacturing, handling, or the on-site movement of such chemicals, or combinations of these activities.

These definitions provide a route to understanding the situations in pilot plants and laboratories for which PSM is most beneficial. For example, a biologic lab making 1000 L of a physiological buffer, which requires pH adjustment using a concentrated acid or base, is not using a "highly hazardous chemical" as defined by PSM. An organic synthesis lab using 10 – 100 mL quantities of a pyrophor does meet this definition. Similarly, a NMR instrument does not meet the definition of a "process" while an HPLC does, because of the highly hazardous chemicals it utilizes.

To understand how MOC should be applied, consider the following incident. A pilot plant was set up to test catalysts in a non-flammable solvent system. The feedstock was phosphorous oxychloride, which has a PSM threshold of 1000 lbs. The unit had a 50 gallon supply or 625 pounds of POCl. Thus, the OSHA PSM standard did not apply. Bench scale testing had demonstrated that a new catalyst provided a 30 – 40 percent increase in efficiency. When the new catalyst was run in the unit, the operator had to initiate an emergency intervention to force a shutdown due to excessive heat generation.

If the unit scientists had followed the PSM guidance on MOC, this incident could have been avoided. The MOC guidance has five points;

- The technical basis for the proposed change;
- Impact of change on safety and health;
- Modifications to operating procedures;
- Necessary time period for the change; and,
- Authorization requirements for the proposed change.

Had the unit scientists looked in detail at the second and third bullets, they should have realized that a 30 – 40% increase in efficiency would mean that heat would be generated at about that same increased rate and that additional heat removal capacity may be necessary. The original process hazard analysis did not address heat removal and the scientists were not prepared to consider it. A formal MOC review would have forced heat removal to be considered. The incident investigation cited as a contributing cause the lack of a clear policy on when a Management of Change review is needed. The scientists involved did not understand "when a change is really a change."

MOC Triggers

A process hazard analysis (PHA) must evaluate the risks associated with a set of conditions for a given reaction. Physical parameters such as temperature, pressure, reactant flow rates, stirring, and others are considered. This allows the PHA team to set maximum acceptable values and operating ranges for each parameter. Chemical parameters including the specific chemistry, the allowable rate of reaction, pH, acceptable catalysts and acceptable solvents must be established during the PHA.

A MOC review is then triggered whenever the research goes outside these parameters. Typical triggers include

- Solvent flammability/toxicity and other hazard changes
- pH changes by more than ± 2 (or 3) pH units
- Unit chemistry changes significantly
- Scale-up that greater than or equal to three-fold
- Whenever a significant unit parameter is changed by ±10 – 20%

This approach of defining an acceptable range of parameters to control the risks associated with research can be depicted as in figure 1.

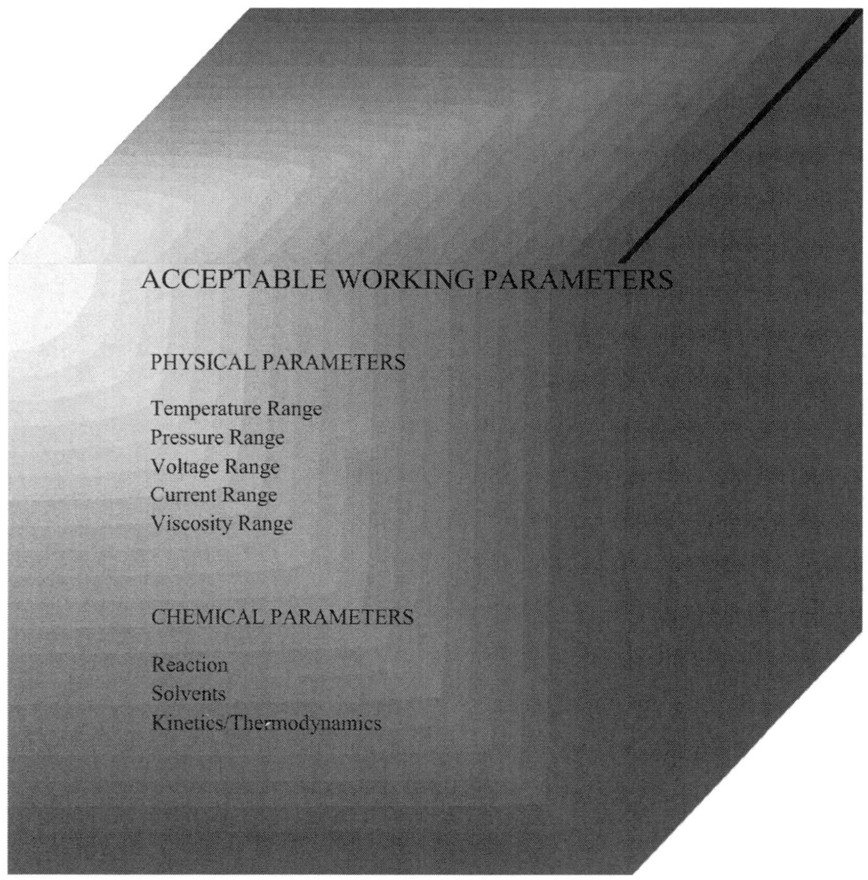

Figure 1. Conceptual field of control of acceptable risk for research parameters

The results of the PHA can be conveniently summarized in tabular form as shown in Table 1. This table can be taped to the unit or fume hood and provides an immediate reminder to the scientists of the range of changes permitted without need of further process hazard analysis.

Table 1

SAFETY CHECKLIST			
Chemistry or Process Set-up			

(insert PHA control parameters)
(insert chemical reaction or other information, as appropriate)

UPSET CONDITION ACTIONS:

HAZARDS

Toxic/High Toxic/Repro		Cooling Required	
Carcinogenic		Max Temp	
Flammable		High Warning Temp	
Alkali Metals		High Action Temp	
Pyrophoric		Pressurized System	
Peroxide Former		Max Press	
Explosive and Explosive formers		High Warning Press	
Oxidizers		High Action Press	
Nanomaterials		Gas(es) Evolved	
Acids/Bases (>10%)		Toxic Gases	
Nitric Acid (>10%)		Hydrogen (Flammable Gases)	
Powders (eg Silica Gel)		Distillation Temp	
HF		Flash Chromatography (Temp)	
		Voltage	
Unattended Operations		Dry Ice	
Operator is:		Gas Cylinders	
Contractor		Heat	
Student		Vacuum System: range in torr	
		Cryogenic	
		Sharps	
		Needles	
Reaction is Literature Prep (Y/N)			
Reaction has been recently performed at this scale without incident (Y/N)			
Reaction is scale-up			
Scale-up Ratio			

Summary

Research in the chemical laboratory or pilot plant is all about continual change. While incremental changes in reaction conditions are usually benign, some changes can trigger an upset resulting in a fire or explosion. Thus, formal management of change should be part of a normal research protocol. In a culture of safety this can be readily implemented by including a brief discussion of the safety implications of any proposed experiment in the regular meetings of the research group. The PHA summary in Table 1 can be used during such a discussion to help determine if a new or uncontrolled hazard is being introduced. This approach can not only help prevent an undesired incident but the frequency of including such safety discussions is a leading metric which can be used to further control research risks.

The OSHA Process Safety Management regulation is a useful guide to risk management in the research environment. Although the regulation usually does not apply to pilot plants or laboratories, it can still be useful as a guide, without the need for a formal PSM program. Application to the research setting is easy and does not impart onerous time constraints on the research team.

References

1.   https://www.osha.gov/pls/oshaweb/owadisp.show_document?p_table= STANDARDS&p_id=9760 (site tested 9 October 2013).
2.   This chapter is an extension of published previous work. Langerman, N. Management of Change for Laboratories and Pilot Plants. *Org. Process Res. Dev.* **2008**, *12* (6), 1305–1306. Langerman, N. Lab-Scale Process Safety Management. *J. Chem. Health Safety* **2009**, *16* (4, July–August), 22–28.

Chapter 5

Transitioning Culture: Teaching and Modeling Workplace Behavior

M. A. Thomson* and W. Killian

Physical Sciences Department, Ferris State University, 820 Campus Drive, Big Rapids, Michigan 49307, United States
***E-mail: MarkThomson@ferris.edu**

As part of a well-established Industrial Chemical Technology program at Ferris State University, two particular courses have sought to introduce students to the requirements and expectations of the workforce. The first is a course in Safety in the Chemical Laboratory. The second is a Chemical Manufacturing and Analysis lab class. These courses are discussed both in the context of preparing students for the culture of the workforce and feedback and assessment from employers.

Introduction

As defined by Webster's New Collegiate Dictionary, culture is "the integrated pattern of human behavior that includes thought, speech, action, and artifacts and depends on man's capacity for learning and transmitting knowledge to succeeding generations." Chemistry, as a significant and important part of the broader culture of science, is the systematic study of material and how it changes. This systematic study is accomplished in two different and distinct sub-cultures, Academia and Industry. Both of these sub-cultures have developed very different patterns of behavior as reflected in their value hierarchy. They have also developed different strategies for transmitting knowledge and behavioral expectations to subsequent generations. Transition between the two sub-cultures can be problematic if the transitioning individuals are unaware of these differences or are unprepared for such transitions.

The Industrial Chemical Technology (ICT) Program at Ferris State University (FSU) is a two year Associate of Applied Science (AAS) Degree program. In this program, some excellent methods have been found to enhance the teaching, learning, and assessment so as to better facilitate the transition of students into the sub-culture of the Industry workforce. Many of these methods were developed by individual faculty members who were also transitioning to full time positions in the Academia sub-culture after spending considerable time in Industrial positions. Two primary areas which will be focused on here have risen to prominence in this FSU-ICT Program, differentiating it from many other Chemistry programs. These areas are Laboratory Safety and Laboratory Time Management, both of which will help to illustrate differences between the two sub-cultures that make the transitions and expectations challenging.

Establishing a Culture of Safety

Chemistry laboratories are areas full of potential hazards at almost every level. Inventories include substances that might be toxic, carcinogenic, or mutagenic as well as substances whose hazards might not be known. Furthermore, these substances need to be stored and such storage needs to accommodate certain substance incompatibilities. Reactions can also involve very large changes in pressure, heat, and volume, creating additional hazards that are not always anticipated, especially when changing the scale of the reaction. Recognition of these hazards is demonstrated by the extent of governmental regulations that must be followed when operating a chemical enterprise.

Safety in Academia

The sub-culture established in Academia reflects the segmented nature of the instruction schedule, the length of time designated for all lab tasks and instruction, the amount of academic credit given, and the other additional outside tasks that also compete for student time and attention. Within these constraints, academic departments and faculty members must find the best possible solution for preparing students for a variety of future expectations.

In a typical undergraduate chemistry course, safety instruction is provided during the first week of each new lab course each semester. This is done because safe behavior expectations must be established before any other work or instruction can begin. Unfortunately, this is also a period of significant student enrollment flux and change as students modify their course load at the onset of each semester. While this is done for sound administrative and academic reasons, the result is a student cultural attitude that the first week of the semester is unimportant and expendable.

Future safety instruction throughout the semester is often given strictly on a "need to know" basis. It may come as a bullet list in a lab manual or as a small portion of a pre-lab lecture/discussion. In both cases, it can appear as a brief after thought when the primary student interest and focus is "I need to get started quickly so I can finish early and leave because I have other things to do." Safety is rarely a part of student evaluation or assessment and the students and faculty both recognize that a student would never fail a course solely due to unsafe practices.

The reasons for faculty attitude toward safety may be reflective of the reward system and value hierarchy of the culture. Top faculty priorities are typically centered on tenure and promotion. At most institutions, these are based on either a publication and funding record or on student performance and evaluation. Safe behavior in lab, while important, can be seen as time consuming and as something that is neither rewarded nor punished.

Safety in Industry

Behavior and culture in Industry also follows well-established traditions. Competing interests include company profitability (as dictated by the stakeholders), worker/workplace conditions (as dictated by labor organizations), and community/environmental sustainability (as dictated by governmental entities). The interests have been resolved and continue to be negotiated through a complex series of regulations and agreements. The end result is a collective, well-monitored and regulated workplace. Safety becomes the highest priority because it is a communal concern and because unsafe practices result in work stoppages that eliminate the possibility of accomplishing any other objectives. The resulting investigations that are precipitated by both safety incidents and near incidents can be incredibly expensive and time consuming. Companies that do not take appropriate measures will not be able to remain profitable.

Safety becomes a communal concern and a collective responsibility. The workspace is no longer an independent area where activities are concentrated on achieving a personal goal or grade. The rewards of the group's efforts result in team or company success and profitability. The space is shared and the activities are expected to continue in production after individuals either leave for the day or find employment elsewhere. Maintaining a safe environment to work in becomes an issue of sustainability and longevity rather than just a temporary road to a grade at the end of a brief semester or lab experiment.

In addition to communal concerns, safety in Industry becomes a regulatory concern. Because of differences in scope and scale, lab practices are regulated differently in an Industry setting. Failure to behave appropriately can bring about work stoppages and shutdowns. With greater external consequences and their implications come much more internal effort to comply and avoid such consequences. Safe practices, hazard prevention, and threat avoidance become the object of more training and priority. Safety becomes Job One.

Safety in the ICT Program at FSU

A course was established early in the development of the ICT Program at FSU to address safety issues, especially to prepare students for Industry expectations. This course, CHEM145 Chemical Lab Safety, is a 2 hour lecture course taught during the second semester of the first year of study and serves to focus the attention of students on all topics related to safety. At this point, they have already completed a semester of General Chemistry with the associated three hour per week lab exercises so they have concrete experiences to contextualize the discussion. Class size is kept small so that lab-based hands-on activities can be included to further enrich the topics under discussion. For example, the entire group can relocate to an available lab space to view practical aspects of chemical storage and compatibility, demonstrate and practice the use of different types of fire extinguishers or observe and discuss the operation of hoods. Topics covered in the course include but are not limited to physical and chemical hazards, storage compatibility and MSDS sheets, corrosives, combustibles, explosives, radiation, and general lab safe practices. These topics are usually beyond the scope of traditional instructional lab needs and this course allows for greater depth and understanding by the students.

While 30 hours of safety instruction is not likely to be sufficient time to cover every possible safety issue and concern that will occur in the future career of each student, it provides a sound foundation apart from the distractions that occur while trying to teach and learn other content. Student interest and attention follows wherever the instructor or institution places importance and gives credit. Students realize that they must pass this course in order to progress in their education and obtain their degree. For that reason, they pay attention to the safety topics presented and are concerned about demonstrating, through graded assessments, that they understand the best practices and will follow them.

Establishing a Culture of Time Management

Distribution of time and time management issues are some of the most difficult aspects that students face as they transition from High School to College. They are moving from a highly structured environment where their time is firmly scheduled for 7-9 hours daily. Within this time span, they have tasks to accomplish and a rigid time frame for accomplishing them. If they are involved in additional extra-curricular activities, this structured time period might extend for several more hours every day.

As these students transition to college, their lifestyle with respect to scheduling and time management shifts dramatically. Now less than 20 hours per week are occupied by specific, structured activity. This time is often sporadically distributed through the week and might include lecture time, lab time, and workshop time. Additional, there is an expectation that they will spend 20-40 hours more on their work at other times on a weekly basis. As every college advisor know, this has the potential for serious consequences experienced by many students who are unable to make the necessary adjustments.

Time Management in Academia

Academic culture has developed a very segmented and piece meal approach to time management. The building blocks for schedule development are small, on the order of minutes when considering lectures and on the order of just a few hours when thinking about laboratories. Because the geographic resources of classroom and laboratory space are shared and used by a very large number of students, time with that resource is dispersed on an "as needed" basis to maximize efficiency. A single laboratory room might be shared by as many as 75 students daily, occupying the space in 175 minutes (3 hours blocks). Tasks and projects are designed to attend the needs of minimally competent students, time required for instructions, discussions, and explanations of safety concerns.

Each semester (15 weeks) course has a broad range of lecture topics that need to be covered and reinforced through laboratory experiments. Specific skills need to also be developed, improved, and refined. This volume of material typically requires that each experiment be taught in a single laboratory period without carrying over for more than one week. Long term laboratory issues such as safety, inventory, and waste management/disposal are shortened or even eliminated to leave room for more important content issues. Knowledge acquisition and assessment of that knowledge become the driving force for effort spent.

Academic culture further constrains experience by scheduling instruction and lab work in two semesters per year. This gives students the perspective that "working full time" means completing 24-30 projects per year, each of which can be finished in just over two hours. These projects are also most often completed individually or with a single partner.

Time Management in Industry

Industry culture is much more collective. Projects are not simple 2-3 hour exercises that are completed and filed away until next week. Planning and time management make use of a much larger building block, thinking of the 40 hour week as the basic scheduling unit. Projects are measured in terms of weeks or months rather than minutes or hours.

Longer term projects require longer term planning. Collective projects and shift work also involve continuity of project rather than completion. At the end of an 8 hour day or shift, the current set of experiments often remains unfinished. It must either be handed off to the next shift or safely put on hold until tomorrow morning. While the 8 hour day can be adjusted with flex-time in some instances, there is still an expectation of spending much more than 2-3 hours of consecutive time in a laboratory setting each week.

Industry culture also requires managing and planning for multiple and competing time constraints. Tests need to be performed, meetings need to be attended, and reports need to be complied and filed. Often these tasks all need to be completed simultaneously. This requires a much different skill set than the one conducive to success in an academic culture.

Time Management in Transition – Ferris State University

The ICT program at Ferris State University was originally established in 1957 with a goal to prepare graduating high school students to be workforce ready for the chemical industry. It has continued in this tradition in a fashion similar to many such programs across the country which continue to prepare Applied Chemical Technical Professionals (ACTPs). Consequently, the FSU program has always had a strong presence of faculty with industrial experience and relied on the support and guidance of an advisory board made up of industry-based partners.

From its inception, the ICT program has developed non-traditional course work, like the safety course described above, to prepare students to an Industry culture. Another such course is CHEM 245 – Chemical Manufacturing & Applied Analysis. This is an eight hour lab course that meets once per week in an effort to simulate a typical work day. It is typically taught during the spring of the final year in the program as students are preparing to graduate.

Chemical Manufacturing is a course with a variety of features to help students adjust to the upcoming time management changes they are about to encounter. Each week's lab starts with a one hour project meeting. Goals, projects, and planning for the day are presented and discussed at this mandatory meeting. The remainder of the time spent in lab during the day remains somewhat flexible. The instructor recognizes that student schedules might require enrollment in other classes with conflicting schedules. Each student, as part of a project team, is expected to negotiate time management with other team members for breaks to attend other classes or eat lunch. Through this negotiated effort, the project continues and the tasks are accomplished even though each team member is in lab on a different schedule. The team must collectively carry the responsibility of completing the reactions and analysis in a timely and safe fashion. Evaluation at the end of the semester is based on exams, a lab notebook, and peer presentations to the students in the course. Feedback from recent graduates, their employers, and advisory board members continues to highlight this course as an important and vital asset in the program. It has made transitions easier and has been described as beneficial, even by those students completing this course as a supplement to a BA degree in Chemistry and progressing to graduate school and PhD studies.

Conclusion

To borrow from Diogenes, "There is nothing permanent except change." Students, throughout their career will constantly face change and transition. Transition between levels of education, between education and the workforce, and within the workforce between different positions, projects, and employers. These cultural transitions, especially for the inexperienced, can be rather challenging and stressful. With planning and assistance, potential disasters can be avoided and replaced with positive growth experiences to build on for a lifetime. Important in the process is the recognition of these transitions and a willingness to think about creative, non-traditional solutions. These solutions worked to the benefit of students, faculty, and graduates at Ferris State University and will continue to do so in the future as long as change is welcomed and embraced.

Chapter 6

Blast Energy and Damage Assessment from the Mechanical Explosion of Compressed Gas Cylinders

Elmer B. Ledesma*

University of St. Thomas, Department of Chemistry and Physics, Houston, Texas 77006
*Tel.: +1-713-831-7810; E-mail: ledesme@stthom.edu

Gas cylinders are common in industrial, academic, and medical environments. Among the hazards associated with the handling and operation of gas cylinders, the most important is the considerable amount of compressed energy stored in the high pressure gas. Rupture of a gas cylinder can result in a mechanical explosion, releasing the compressed energy of the gas. This chapter demonstrates how to estimate the blast energy and assess the damage resulting from the explosion process of a compressed gas cylinder. In this chapter a 300 ft^3 compressed helium cylinder at 77°F and at 1450, 2900, and 5800 psi initial pressures, is used as an example. The analysis of the explosion process is achieved by constructing mass, energy and entropy balances for the system. The resulting thermodynamic equations are solved for the blast energy using three methods: (1) using available experimental thermodynamic data for the gas; (2) using the Redlich-Kwong generalized equation of state; and (3) assuming that the gas behaves ideally. Assessment of the damage, based on overpressures, from the corresponding blast energy is discussed.

Introduction

Gas cylinders containing gases under high pressure are very common in chemical plants, academic and medical laboratories. The compressed gas or gases stored in these cylinders are used in a variety of applications such as a reactant in synthetic procedures, a blanketing agent to prevent undesired chemical reactions, or as a carrier gas in chromatographic separations. There are a wide variety of sizes of gas cylinders that are commonly used in chemical plants and laboratories. A common size used in laboratories is 9 in diameter by 51 in high and an approximate volume of 300 ft^3. Due to their ubiquity in industrial and academic settings, their proper handling and storage are important, as gas cylinders present particular hazards that are not commonly encountered when handling chemicals in the liquid and solid forms.

The main hazards associated with the use of gas cylinders are the following: (1) cylinder valve or regulator malfunction, potentially resulting in the cylinder becoming a projectile that will destroy everything in its path; (2) high pressure discharge leading to a mechanical explosion; and (3) gas leak due to a faulty cylinder valve or regulator. A leaking cylinder may have multiple hazards: (a) inert gases can reduce the concentration of oxygen in the air in its vicinity which could lead asphyxiation; (b) flammable gases can result in fire or explosion; (c) oxygen leaks can support rapid combustion, even with materials not ordinarily regarded as readily flammable; (d) poisonous gases can lead to health hazards or death; and (e) corrosive gases can result in equipment damage and failure elsewhere in the pilot plant with the potential for chemical exposure and injury.

One of the most important hazards associated with gas cylinders is the high pressure of the compressed gas. Commercial gas cylinders are typically filled at pressures between 2200 to 2500 psi. However, in pilot plant operations gas mixtures and higher pressures are often used and their containment cylinders must be carefully assessed for their hazard potential. Overpressurization increases the stress on the cylinder, thereby rupturing the cylinder, and a mechanical explosion ensues as the gas rapidly expands, releasing its energy of compression.

Due to the explosive energy contained in gas cylinders, mechanical explosions resulting from the sudden discharge of the high pressure gas is the main focus of this chapter. The important safety variable to determine is the blast energy. In order to estimate the blast energy of a mechanical explosion, a thermodynamic analysis of the process is described and discussed to arrive at equations for calculating the blast energy. As an example in using the thermodynamic equations to estimate blast energy, the mechanical explosion of a 300 ft^3 compressed helium cylinder is presented. Helium is commonplace in academic, industrial, commercial and medical settings. It is used as a carrier gas for gas chromatography, a shielding gas in arc welding, a cooling gas in cryogenic applications, protective gas in the synthesis of some noble metals, and a lifting gas in recreational balloons and commercial airships.

Thermodynamic Analysis of Mechanical Explosions

Overview

A mechanical explosion is a rapid, uniform expansion of a gas that takes place without chemical reaction. The process is very rapid such that heat and mass transfer to or from the expanding gas can be considered negligible. A shock wave is produced where the pressure inside the wave is much greater than the ambient pressure outside the wave. The shock wave continually travels outward until the pressure inside equalizes to that of the ambient pressure. The speed of the shock wave depends on the initial pressure. An *explosion* results if the initial pressure is high such that the shock wave travels at the speed of sound. A *deflagration* occurs if the wave travels slower than the speed of sound. The damage imparted by an explosion is primarily due to the release of energy in the shock wave.

To estimate the blast energy for the explosion of a compressed gas cylinder, a thermodynamic analysis is performed. The system is taken to be the gas in the cylinder at an initial temperature of T_i and an initial pressure of P_i. The initial pressure is also the pressure at which the cylinder ruptures, causing an explosion. The system is a closed system whose boundary is expanding. Application of the conservation of mass leads to the mass balance equation,

$$N_f = N_i \tag{1}$$

where N_i and N_f are the initial moles of gas before the explosion and the final moles of gas once the pressure inside the shock wave is at the ambient pressure of P_f, respectively. Since the expansion is very rapid and occurs uniformly, the system is considered adiabatic without any temperature or pressure gradients, with the exception of the system boundary.

Application of the conservation of energy for a closed system gives the following energy balance equation,

$$N\left(\underline{U}_f - \underline{U}_i\right) = W \tag{2}$$

where $\underline{U}_i$ and $\underline{U}_f$ are the initial and final molar internal energies, N is the moles of gas present in the system (this is determined from the initial conditions, which according to the mass balance is also equal to the final moles), and W is the work transferred. W is also the blast energy.

The molar internal energy terms in Equation (2) are functions of temperature if the gas behaves ideally or functions of both temperature and pressure if the gas is considered a real gas. In either case, to compute the blast energy, the final temperature of the system, T_f, is required. This is unknown. In order to determine W from Equation (2), T_f must be computed using another equation. This additional equation is obtained from the conservation of entropy. In addition to being a rapid, uniform expansion, the explosion is also assumed to be isentropic with neglect of the generation of entropy. This is an approximation as it is difficult, if not impossible, to compute the generation of entropy during an explosion. It should be noted that neglect of the entropy generation term will result in a blast energy value that is an upper bound to the true value. This is in fact desired, for in safety problems involving explosions, it is always best to be conservative

As a result of neglecting the entropy generation term, the entropy balance equation is given by,

$$\underline{S}_f - \underline{S}_i = 0 \tag{3}$$

where $\underline{S}_i$ and $\underline{S}_f$ are the initial and final molar entropies. This equation is used to determine the unknown final temperature. Once T_f is determined from Equation

(3), the blast energy can then be computed using Equation (2). Blast energy is commonly expressed as the mass of trinitrotoluene (TNT) that results in an equivalent amount of energy released, where the blast energy of TNT is 1980 BTU per lb.

In order to use the thermodynamic equations to calculate blast energy, experimental thermodynamic data or volumetric equations of state to represent the pressure-volume-temperature (PVT) behavior of the gas are needed. In this study, experimental data and volumetric equations of state are used.

Ideal Gas

If the gas behaves ideally, molar internal energy is a function of temperature only. Therefore, the energy balance given in Equations (2) for an ideal gas can be expressed as,

$$N \int_{T_i}^{T_f} (C_P - R)\, dT = W \tag{4}$$

where T_i is the initial temperature, C_P is the ideal-gas, constant-pressure heat capacity and R is the gas constant. For an ideal gas, entropy is a function of both temperature and pressure, and the entropy balance given in Equation (3) can be expressed by the following,

$$\int_{T_i}^{T_f} \frac{C_P - R}{T}\, dT - \int_{P_i}^{P_f} \frac{R}{P}\, dP = 0 \tag{5}$$

where P_i and P_f are the initial and final pressures, respectively.

Depending on the gas, the constant-pressure heat capacity term in Equations (4) and (5) is either a constant (common for monatomic gases) or is temperature-dependent, customarily expressed as a cubic polynomial in terms of the temperature, T:

$$C_P = a + bT + cT^2 + dT^3 \tag{6}$$

The coefficients in the cubic polynomial for C_P are well tabulated in the literature for many engineering fluids (*1–6*). Due to the mathematical form of C_P, the integrals in Equations (4) and (5) are elementary, and analytical expressions for the energy and entropy balances can be derived. For constant C_P, the equations are as follows

$$NC_P(T_f - T_i) = W \tag{7}$$

$$C_P \log_e \frac{T_f}{T_i} - R \log_e \frac{P_f}{P_i} = 0 \tag{8}$$

Therefore, in order to estimate the blast energy W upon explosion of a cylinder gas cylinder (assuming ideal gas behavior) at initial temperature and pressure, T_i and

P_i, respectively, and with an ambient pressure of P_f, Equation (8) is first used to determine T_f. Once T_f has been evaluated, it is then substituted into Equation (7) to obtain W.

Generalized Equation of State

If the gas is considered real, the molar internal energy and entropy are functions of both temperature and pressure. As a consequence, expressions for changes in molar internal energy, $\Delta \underline{U}$, and entropy, $\Delta \underline{S}$, in going from the an initial state (T_i, P_i) to a final state (T_f, P_f) are more complicated for a real gas than those for a gas that is considered ideal. For real gases, these expressions are as follows,

$$\Delta \underline{U} = \Delta \underline{H}^{IG} + \left(\underline{H} - \underline{H}^{IG} \right)_{T_f, P_f} - \left(\underline{H} - \underline{H}^{IG} \right)_{T_i, P_i} - \Delta(P\underline{V}) \tag{9}$$

$$\Delta \underline{S} = \Delta \underline{S}^{IG} + \left(\underline{S} - \underline{S}^{IG} \right)_{T_f, P_f} - \left(\underline{S} - \underline{S}^{IG} \right)_{T_i, P_i} \tag{10}$$

with

$$\Delta \underline{H}^{IG} = \int_{T_i}^{T_f} C_P \, dT \tag{11}$$

$$\Delta \underline{S}^{IG} = \int_{T_i}^{T_f} \frac{C_P - R}{T} \, dT - \int_{P_i}^{P_f} \frac{R}{P} \, dP \tag{12}$$

$$\Delta(P\underline{V}) = R\left(Z_f T_f - Z_i T_i \right) \tag{13}$$

where Z_i and Z_f are the compressibility factors at (T_i, P_i) and (T_f, P_f), respectively. The compressibility factor, Z, is defined as

$$Z = \frac{P\underline{V}}{RT} \tag{14}$$

where $\underline{V}$ is the molar volume. The $(\underline{H} - \underline{H}^{IG})$ and $(\underline{S} - \underline{S}^{IG})$ terms are referred to as the enthalpy and entropy departure functions, respectively, and are evaluated at both the initial and final states. These departure functions are given by the following expressions,

$$\left(\underline{H} - \underline{H}^{IG} \right)_{T,P} = RT(Z - 1) + \int_{\infty}^{\underline{V}(T,P)} \left[T \left(\frac{\partial P}{\partial T} \right)_{\underline{V}} - P \right] d\underline{V} \tag{15}$$

$$\left(\underline{S} - \underline{S}^{IG} \right)_{T,P} = RT \log_e Z + \int_{\infty}^{\underline{V}(T,P)} \left[\left(\frac{\partial P}{\partial T} \right)_{\underline{V}} - \frac{R}{\underline{V}} \right] d\underline{V} \tag{16}$$

In order to use Equations (9) and (10) to determine changes in internal energy and entropy in going from an initial state to a final state, the integrals need to be evaluated. The first integral on the right hand side in each equation, $\Delta \underline{H}^{IG}$ and $\Delta \underline{S}^{IG}$, is elementary as the C_P term is either a constant or a cubic polynomial in terms of T. The integrals given in the departure functions of Equations (9) and (10) require an equation of state for a real gas.

There have been many generalized equations of state developed to represent real gases. In this study, the Redlich-Kwong (RK) equation of state is used. The RK equation is a commonly used equation of state in industry. It is given by the following expression,

$$P = \frac{RT}{\underline{V} - b} - \frac{a}{T^{1/2}\underline{V}(\underline{V} + b)} \tag{17}$$

where the equation of state constants, a and b, are defined as

$$a = 0.4278 \frac{R^2 T_C^2}{P_C} \tag{18}$$

$$b = 0.08664 \frac{R T_C}{P_C} \tag{19}$$

In Equations (18) and (19), T_C, P_C, and ω are the critical temperature, critical pressure, and accentric factor. Values of these for many engineering fluids are well tabulated in the literature (1–6). The RK equation of state is a cubic equation in terms of $\underline{V}$. It can be equivalently expressed as a cubic polynomial in terms of the compressibility factor as follows,

$$Z^3 - Z^2 + (A - B - B^2)Z - AB = 0 \tag{20}$$

where

$$A = \frac{aP}{R^2 T^{5/2}} \tag{21}$$

$$B = \frac{bP}{RT} \tag{22}$$

Using Equation (17), the enthalpy and entropy departure functions given in Equations (15) and (16), respectively, can be written as,

$$\left(\underline{H} - \underline{H}^{IG}\right)_{T,P} = RT(Z - 1) + \frac{3a}{2bT^{1/2}} \log_e \frac{Z}{Z + B} \tag{23}$$

$$\left(\underline{S} - \underline{S}^{IG}\right)_{T,P} = R \log_e(Z - B) + \frac{a}{2bT^{3/2}} \log_e \left(\frac{Z}{Z + B}\right) \tag{24}$$

Similar to the ideal gas case, to estimate the blast energy W upon explosion of a compressed gas cylinder at initial temperature and pressure, T_i and P_i, respectively, and with an ambient pressure of P_f, the final unknown temperature, T_f, must be determined. This is accomplished by finding the value of T_f and Z that makes the left-hand side of Equation (10) equal to zero and that satisfies Equation (20). Both Equations (10) and (20) are nonlinear and a numerical method is needed to obtain a solution. Once the final temperature has been determined, it is then used to determine the blast energy using Equations (2) and (9).

Experimental Thermodynamic Data

Applying the ideal gas and RK equations of state are useful if experimental thermodynamic data are not available. If available, then the experimental data will provide the most realistic and accurate results. Experimental thermodynamic data for helium are available from the NIST Chemistry WebBook (*7, 8*). The experimental data are generated by an accurate equation of state that is specific for helium (*9*). Experimental data also make the calculations straightforward, as Equation (3) is first satisfied by finding T_f. This is accomplished by finding the initial entropy and then interpolating on a temperature such that the $\underline{S}_f$ equals $\underline{S}_I$. The blast energy is then easily determined from Equation (2).

Mechanical Explosion of Compressed Helium

In order to illustrate the use of the thermodynamic equations derived above, the blast energy resulting from the explosion of a 300 ft³ compressed helium cylinder is determined. With very slight modifications, the same equations are applicable to other gases and gas mixtures as well. It is assumed that the initial temperature of the system is 77°F. Three initial cylinder pressures are investigated: 1450, 2900, and 5800 psi. For a 300 ft³ cylinder at 77°F, 1450 psi approximately represents a half-filled cylinder, 2900 psi approximately represents a completely filled cylinder, and 5800 psi approximately represents an over-filled cylinder. Tables 1-3 present the calculated results for T_f and blast energy.

The results in Tables 1-3 show that the final temperatures derived from the three different methods at every initial pressure are very similar to one another. The blast energies, however, deviate, especially at the higher initial pressures. At initial pressure of 1450 psi, the computed blast energies are similar to one another, with the RK result showing lower deviation (-2.3%) than the ideal gas value (+4.2%) from the experimental result. At 2900 psi, the deviations are -4.3% and +8.5% for the RK and ideal gas, respectively. At the highest initial pressure investigated, the deviations are -7.7% and +16.9% for the RK and ideal gas, respectively. These deviations in blast energy from the experimental results are consequences of the increasing non-ideality of the gas with an increase in the initial pressure. At 1450 psi, the compressibility factor (computed from the experimental thermodynamic data obtained from the NIST Chemistry WebBook (*7, 8*)) is 1.046, increasing to 1.185 at 5800 psi.

The results demonstrate that in order to obtain an accurate result for the blast energy of a mechanical explosion, accurate thermodynamic data are needed. Assuming that the compressed gas is ideal is the simplest method, but it is inaccurate, especially at high initial pressures. The next simplest method is to use the experimental thermodynamic data of the gas. This is also the most accurate. If experimental data are not available, a generalized equation of state, like the RK, to represent the PVT behavior of the gas can be used. As illustrated in Tables 1-3, the RK approach produces results closer to the experimental value than the method using the ideal gas equation of state.

Table 1. Calculated final temperatures and blast energies at 1450 psi initial pressure.

	T_f *(°F)*	*Blast Energy (lb TNT)*
Experimental	-373.9	49.27
Redlich-Kwong	-374.1	48.15
Ideal Gas	-374.1	51.35

Table 2. Calculated final temperatures and blast energies at 2900 psi initial pressure.

	T_f *(°F)*	*Blast Energy (lb TNT)*
Experimental	-394.5	99.03
Redlich-Kwong	-394.8	94.78
Ideal Gas	-394.9	107.4

Table 3. Calculated final temperatures and blast energies at 5800 psi initial pressure.

	T_f *(°F)*	*Blast Energy (lb TNT)*
Experimental	-410.0	189.9
Redlich-Kwong	-410.5	175.2
Ideal Gas	-410.5	222.0

Damage Assessment

The damage that results from a mechanical explosion is in part due to the shock wave that is produced. This shock wave impacts on an object. The severity of the shock wave damage depends on the initial pressure and the distance of the object from the point of explosion. Estimates of damage are made by determining the peak overpressure: the greater the overpressure, the greater the severity of the damage. Table 4 relates values of overpressure to damage both structural and physiological (*10, 11*).

Overpressure is calculated by the following empirically-derived equations (*12*),

$$OP = 164z^{-2.01} \qquad 1 \leq z < 10 \qquad\qquad (25)$$

$$OP = 26.5z^{-1.16} \qquad 10 \leq z < 200 \qquad\qquad (26)$$

where OP is the overpressure (psi). In Equations (25) and (26), z is an empirically-derived scaling law given by the expression,

$$z = \frac{r}{W^{1/3}} \qquad\qquad (27)$$

where r is the horizontal distance (ft) from the point of explosion and W is the blast energy (lb).

Using the blast energy values given in Tables 1-3, overpressures as functions of distance and initial pressures are presented in Figures 1-3. The results in the figures show that for all three calculation methods and at distances less than 100 ft, there are marked differences in overpressures as the initial pressure in the cylinder varies. Taking the experimental case (Figure 1), at an initial pressure of 1450 psi, the overpressure value 30 ft from the explosion is 15.4 psi. According to Table 4, this overpressure could result in serious fatalities and total destruction of buildings. At initial pressures of 2900 and 5800 psi, the overpressure values increase to 24.5 and 37.9 psi, respectively. These overpressure values will also result in serious fatalities and total destruction.

As illustrated in Figures 1-3 and for all calculation methods, the overpressure values decrease as the distances from the explosion increase. In addition, at distances greater than 100 ft, there are minor differences in overpressure values. Considering the results derived from the experimental data (Figure 1), at an initial pressure of 1450 psi the overpressure value 100 ft from the explosion is 1.7 psi. According to Table 4, this overpressure value could result in the partial demolition of houses and the failure of metal and wooden panels. Also, this overpressure value is in the range of laceration damage from debris resulting from the rupture of the cylinder. With an increase in initial pressures to 2900 and 5800 psi, the overpressure values increase to 2.2 and 3.4 psi, respectively. As indicated in Table 4, these overpressure values result in significant structural and physiological damage. At 300 ft from the point of explosion, the overpressure values are 0.47, 0.61, and 0.79 psi at initial pressures of 1450, 2900, and 5800 psi, respectively. At this distance, Table 4 indicates that minor structural damage and glass failure will be the result.

The results discussed above for the experimental case clearly show that the greater the initial pressure in the cylinder before rupture and explosion, the greater the amount of explosive energy and hence, the greater the severity of the damage. Similar trends are also observed for the RK and ideal gas cases. The results in Figures 1-3 in conjunction with Table 4 therefore highlight the risks associated in the operation and handling of a 300 ft³ compressed helium cylinder. Even a cylinder that is half-full (1450 psi) contains a considerable amount of explosive energy, and, depending on the distance from the point of failure and explosion, will result in considerable structural and physiological damage.

Table 4. Relation of overpressure values to severity of damage (*10, 11*).

Overpressure (psi)	Structural and Physiological Damage
0.04	Loud noise. Sonic boom glass failure.
0.15	Typical pressure for glass failure.
0.4	Limited minor structural damage.
0.5-1.0	Large and small windows usually shattered.
0.7	Minor damage to house structures.
1.0	Partial demolition of houses; made uninhabitable.
1.0-2.0	Corrugated metal panels fail and buckle. Housing wood panels blown in.
1.0-8.0	Range for slight to serious laceration injuries from flying glass and other missiles.
2.0	Partial collapse of walls and roofs of houses.
2.0-3.0	Non-reinforced concrete or cinder block walls shattered.
2.4-12.2	Range for 1-90% eardrum rupture among exposed populations.
2.5	50% destruction of home brickwork.
3.0	Steel frame buildings distorted and pulled away from foundation.
5.0	Wooden utility poles snapped.
5.0-7.0	Nearly complete destruction of houses.
7.0	Loaded train cars overturned.
9.0	Loaded train box cars demolished.
10.0	Probable total building destruction.
14.5-29.0	Range for 1-99% fatalities among exposed populations due to direct blast effects.

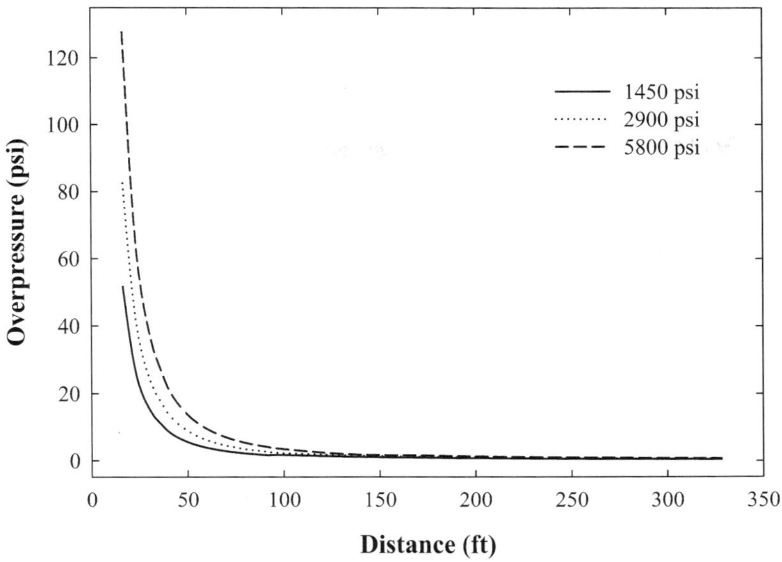

Figure 1. Overpressures as functions of distance curves derived from blast energies computed using the experimental thermodynamic data for helium.

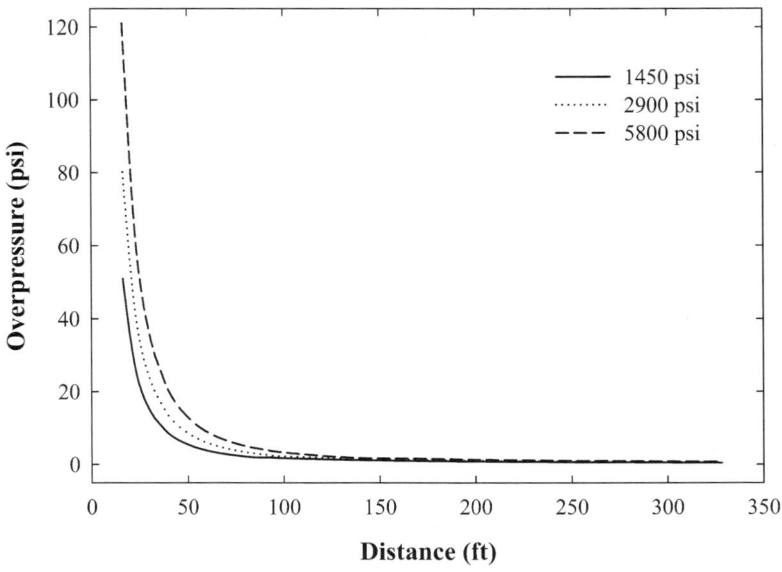

Figure 2. Overpressures as functions of distance curves derived from blast energies computed using the Redlich-Kwong equation of state.

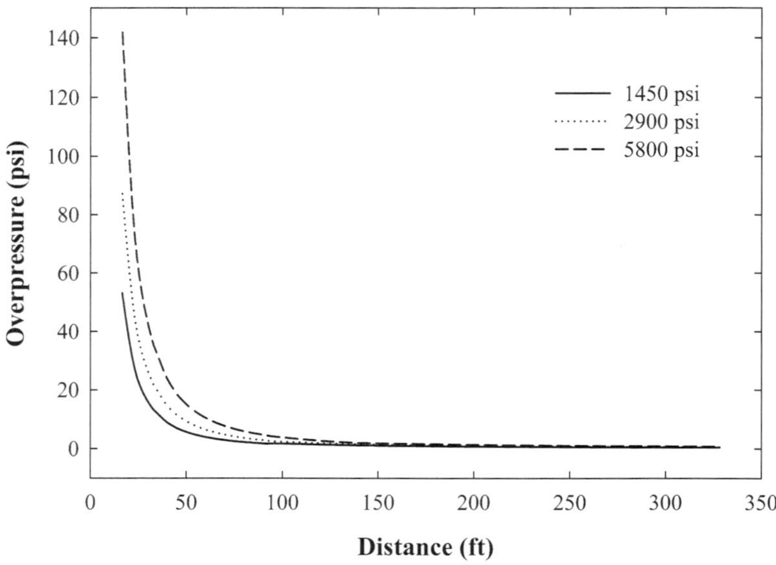

Figure 3. Overpressures as functions of distance curves derived from blast energies computed by assuming helium is an ideal gas.

Conclusions

Blast energies and estimates of the corresponding damage resulting from the mechanical explosion of a 300 ft³ compressed helium cylinder at three initial pressures of 1450, 2900, and 5800 psi, were determined via application of mass, energy and entropy balances on the system. The thermodynamic equations were solved using three different methods: (1) using the experimental thermodynamic data for helium; (2) using the RK generalized equation of state; and (3) assuming helium behaves as an ideal gas. At the lowest initial pressure investigated, blast energies computed from the RK and ideal gas equations of state exhibited small deviations from the experimental value. With an increase in initial pressures, the deviations increased, with the ideal gas result showing the larger deviation from the experimental value than the RK result. The structural and physiological damage associated with the computed blast energies were assessed by determining overpressure values as functions of distances from the point of explosion. For all calculation methods, the results demonstrated that the greater the initial pressure of the cylinder, the greater the severity of the structural and physiological damage. The methodology provided in the chapter provides a means of estimating and understanding the hazard potential associated with compressed gases in the pilot plant environment.

References

1. Perry, R. H.; Green, D. W. *Perry's Chemical Engineers' Handbook*, 7th ed.; McGraw-Hill: New York, NY, 1997.

2. Yaws, C. L. *Chemical Properties Handbook*; McGraw-Hill: New York, NY, 1999.

3. Poling, B. E.; Praisnitz, J. M.; O'Connell, J. P. *The Properties of Gases and Liquids*, 5th ed.; McGraw-Hill: New York, NY, 2001.

4. Sandler, S. I. *Chemical, Biochemical, and Engineering Thermodynamics*, 4th ed.; John Wiley & Sons: New York, NY, 2006.

5. Kyle, B. G. *Chemical and Process Thermodynamics*, 3rd ed.; Prentice Hall: Upper Saddle River, NJ, 1999.

6. Felder, R. M.; Rousseau, R. W. *Elementary Principles of Chemical Processes*, 3rd ed.; John Wiley & Sons: New York, NY, 2005.

7. *NIST Chemistry WebBook*, NIST Standard Reference Database Number 69; Linstrom, P. J., Mallard, W. G., Eds. http://webbook.nist.gov (retrieved April 15, 2014).

8. Lemmon, E. W.; McLinden, M. O.; Friend, D. G. Thermophysical Properties of Fluid Systems. In *NIST Chemistry WebBook*, NIST Standard Reference Database Number 69; Linstrom, P. J., Mallard, W. G., Eds.; National Institute of Standards and Technology: Gaithersburg, MD. http://webbook.nist.gov (retrieved April 15, 2014).

9. McCarty, R. D.; Arp, V. D. *Adv. Cryog. Eng.* **1990**, *35*, 1465–1475.

10. Clancey, V. J. Diagnostic Features of Explosion Damage. Presented at the Sixth International Meeting of Forensic Sciences, Edinburgh, 1972.

11. Crowl, D. A.; Louvar, J. F. *Chemical Process Safety: Fundamentals with Applications*; Prentice Hall: Englewood Cliffs, NJ, 1990.

12. Alonso, F. D.; Ferradás, E. G.; Pérez, J. F. S.; Aznar, A. M.; Gimeno, J. R.; Alonso, J. M. *J. Loss Prevent. Process Ind.* **2006**, *19*, 724–728.

Chapter 7

A Practical Method of Neutralizing Cr(VI) in Phillips Polymerization Catalysts

Kathy S. Collins* and Max P. McDaniel

Chevron-Phillips Chemical Co., Research Dept., Bartlesville, Oklahoma
***E-mail: ColliKS@cphem.com**

The Phillips Cr/silica catalyst, which has great worldwide industrial importance, also involves Cr(VI) at one stage in its preparation. Consequently workers must be protected from exposure and strict decontamination procedures are required. In this paper various ways of neutralizing the Cr(VI) in the catalyst are evaluated with regard to efficiency of reduction, cost and other practical considerations. One reducing agent seems especially well suited to this application, ascorbate (vitamin C). A method of detection of ppm levels of hexavalent chromium on work surfaces is also discussed.

Introduction

Polyethylene (PE) is the most widely used plastic. It is ubiquitous throughout the world economy, used in hundreds of applications, especially in containment and packaging. High density PE (HDPE) is the most widely used form of PE, and about 40% of this material is produced from the Phillips Cr/silica catalyst (*1*). As part of the catalyst preparation it passes through the hexavalent state, so that potential exposure, spills and catalyst waste must be addressed. This is particularly important because hexavalent chromium is a carcinogenic substance, with strong exposure and environmental restrictions by the EPA, OSHA and FDA.

In this paper we report on the development of a new procedure to neutralize Cr(VI) in the Phillips catalyst. This is now used worldwide in many PE manufacturing plants to counteract occasional catalyst spills, or in the decontamination and cleanup of work areas where the catalyst is produced and transferred, and as a general prophylactic against potential worker exposure.

It is well known that Cr(VI) can be converted to the more benign Cr(III) by a number of common organic and inorganic reducing agents in aqueous solution (*2–15*). However, most of these methods are either ineffective or inappropriate for catalyst in a PE manufacturing plant. In this report a number of such reductants have been tested with typical Phillips Cr(VI)/silica catalysts under varied pH conditions. Their reactivity toward hexavalent chromium is described, as well as other advantages and problems associated with these compounds. The most widely accepted treatment, i.e. using ferrous ions, was found to be unreliable except under very acidic conditions. However, ascorbic acid was identified as a particularly potent agent under all conditions, as well as being clean and low in cost.

Results and Discussion

Regulatory Background

The Phillips Cr(VI)/silica catalyst is a powder containing only 1 wt% Cr and with an average particle size of about 100 μm. Thus, when poured or handled inappropriately it can generate dust. Worker exposure limits to air born Cr(VI) dust are quite strict (*16*). OSHA has set legal limits for hexavalent chromium of 0.005 mg/m3 in air averaged over an 8-hour work day and for trivalent chromium of 0.5 mg/m3 in air averaged over an 8-hour work day. NIOSH recommends an exposure limit of 0.5 mg/m3 in air for chromium as chromium metal and divalent and trivalent chromium averaged over an 8-hour work day. NIOSH also recommends an exposure limit of 0.001 mg/m3 in air for hexavalent chromium compounds in air averaged over a 10-hour work day.

Similar strict standards also apply to ingestion of Cr(VI) and to environmental protection (*16*). The EPA has set the maximum contaminate level for total chromium at 0.1 mg/L in drinking water. The FDA determined total chromium concentration should be below 0.1 mg/L in bottled water.

Search for a Reducing Agent

Because of these strong restrictions, it is imperative that the polyethylene plant environment be rigidly kept clean from contamination and worker exposure. To this end, a survey was made of potential reducing agents for the Cr(VI)/silica catalyst, that could be used to neutralize catalyst spills and clean operation surfaces, such as floors, countertops, equipment, railings, sump ponds, and even worker shoes and clothing.

A suitable reducing agent for Cr(VI)/silica catalyst should have the following attributes.

- It should be a strong reducing agent for Cr(VI), neutralizing it quantitatively even at trace concentrations.
- It should be effective at all pH ranges, since strongly acidic or caustic solutions cannot be used on some surfaces or equipment.

- It should be highly water soluble, to provide high concentrations, because the catalyst porosity can adsorb only a limited amount of liquid.
- The reaction products should leave no undesirable stains or sludge behind.
- It should be nontoxic, so that it can be widely used on shoes, floors, ponds that drain to the sewer, etc.
- It should have a low cost.
- It should be reasonably stable over time.

Reduction by Ferrous Ion

The most common general agent used to reduce Cr(VI) is Fe(II), usually ferrous sulfate or ferrous ammonium sulfate in aqueous solution (*17–19*). Ferrous ammonium sulfate has long been used in the PE industry as an analytical test method to determine the amount of Cr(VI) on an activated Phillips catalyst (*1*). A sample of the catalyst that contains Cr(VI) is added to a strong sulfuric acid solution, which leaches the Cr(VI) off the silica surface and into solution. This solution is then titrated with a known concentration of ferrous ammonium sulfate, using phenanthroline as the indicator. Fe(II) is oxidized to Fe(III) by Cr(VI) (*20*). The phenanthroline indicator changes from blue-green in the presence of Fe(III) to red when excess Fe(II) first appears in the solution. The color change is shown in Figure 1.

Figure 1. Titration of Cr(VI) with Fe(II) to the phenanthroline endpoint under strongly acidic conditions. Phenanthroline is blue-green in the presence of Fe(III) and Cr(VI), but a red chelate forms from excess Fe(II) as when the Cr(VI) is fully reduced. (see color insert)

The conversion of Cr(VI) to Cr(III) in this procedure is quantitative because in such strongly acidic solutions the oxidizing power of Cr(VI) is at its greatest potency (*21*). This is indicated in Figure 2 which shows the electromotive potential for Cr(VI) and Fe(II) as a function of pH. Unfortunately, it is not possible to apply such acidic Fe(II) solutions to manufacturing equipment or onto workers or their clothes.

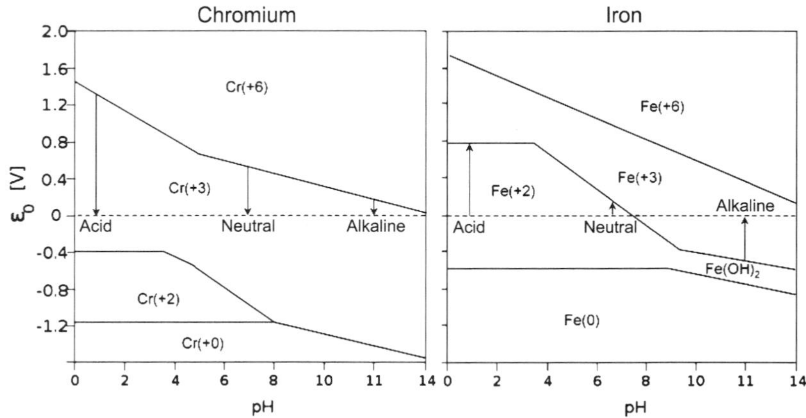

Figure 2. Electromotive potential for the reduction of Cr(VI) by Fe(II), indicating that the reaction is weak or disfavored at high pH.

However, the same reaction has also been widely used under weakly acidic or neutral conditions to neutralize Cr(VI) in ground water (*2–4, 8, 21*). This raises the question of whether a neutral Fe(II) solution could also be used to reduce Cr(VI)/silica polymerization catalysts.

Therefore the ability of ferrous ammonium sulfate to reduce Cr(VI)/silica polymerization catalyst was investigated at neutral pH. Figure 2 predicts that the reaction will indeed take place, although with considerably less force than at strongly acidic pH. Figure 3 shows the result when a few drops of red Fe(II) phenanthroline complex, which serves as the titration indicator in Figure 1, was added to a solution of sodium chromate. In Figure 1 it was immediately converted to the blue-green color of Fe(III) phenanthroline, indicating that the Cr(VI) had oxidized the Fe(II) complex. In this case, however, no color change was observed within the time scale of the experiment (about 1 hour). That is, both Fe(II) and Cr(VI) coexisted in the same solution without reaction. Thus, at neutral pH with the phenanthroline chelate no redox reaction took place. Others have also reported that chelating organic compounds can have an influence on the reaction (*22*). This could be a problem in a polyethylene manufacturing plant, since the ground, other surfaces and sump ponds can have chelating bio-organic matter.

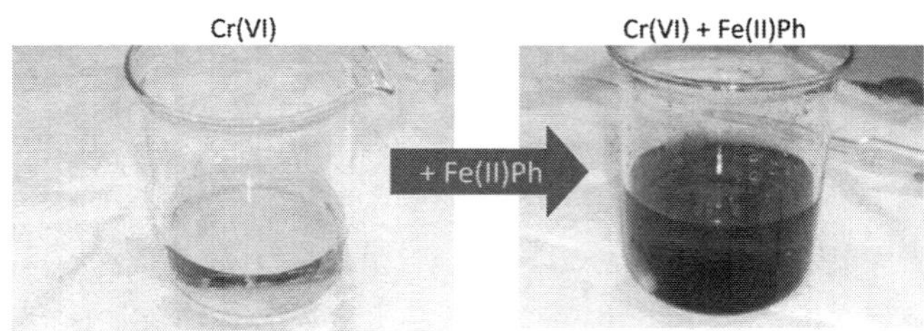

Figure 3. The red Fe(II) phenanthrolene complex remains red when added to a neutral Cr(VI) solution, indicating that Cr(VI) is not neutralized by chelated Fe(II). (see color insert)

The reaction was also tested in alkaline solution. A large excess of ferrous sulfate was added to a sodium chromate solution at high pH. A brownish precipitate immediately formed, probably $Fe(OH)_2$, To avoid the chelating influence on Fe(II), a different indicator was used in this case, diphenyl carbazide, which detects the presence of Cr(VI). When a few drops were added, the suspension immediately turned a red-purple, indicating the presence of Cr(VI). This is shown in Figure 4. Thus, again, Fe(II) did not reduce the Cr(VI) under alkaline conditions, and in the desired time scale. Since alkaline pH might also be used in connection with PE manufacture, for example to encase catalyst in concrete for disposal, this experiment indicates that Fe(II) sulfate would not be the best choice of reducing agent.

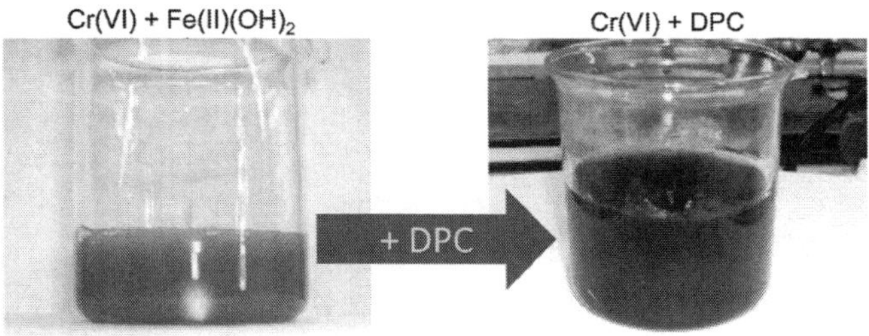

Figure 4. Fe(II) under alkaline conditions forms a yellow-brown precipitate, probably Fe(OH)₂. Adding Cr(VI), then diphenyl carbazide (DPC) produces a dark cloudy red color, indicating that much of the Cr(VI) was not neutralized. (see color insert)

Common Organics as Reducing Agents

It has often been observed that Cr(VI)/silica catalysts, when exposed to a plant atmosphere for a day, are reduced to Cr(III). Sunlight may play a role (*23*), but the reduction has also been observed in the shade. Since plant environments sometimes contain traces of organic matter, which have been reported to slowly reduce Cr(VI) in soils, several organics were tested as reducing agents in solutions at neutral pH. These tests included sodium chromate solution to which was added an excess either of isopropanol, acetone, glycolic acid, oxalic acid, or acetic acid.

Others have reported such natural compounds are effective as reducing agents over several days (*6, 24–28*). Figure 5 shows an example in which a large excess of isopropanol was added to a sodium chromate solution. No change in color occurred within about one hour, indicating that there was no reduction to Cr(III). The conclusion then is that these materials, although capable of slowly reducing Cr(VI) at least partially, are probably too slow to be of use in a polyethylene manufacturing plant.

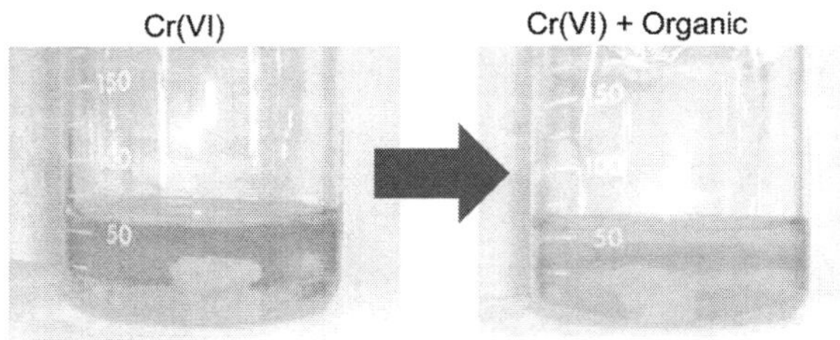

Figure 5. Organic compounds, including isopropanol, acetone, glycolic acid, oxalic acid or acetic acid, were added to chromate solutions at near neutral pH. No reduction of Cr(VI) occurred within an hour. (see color insert)

Sulfites

Another class of reducing agent often cited for neutralization of Cr(VI) are sulfites (*2, 3, 21*). Therefore a solution of sodium bisulfite was tested with Cr(VI)/silica catalysts. Figure 6 shows the result of adding sodium bisulfite to sodium chromate solution. There was an immediate color change from yellow to green. This solution was only mildly acidic.

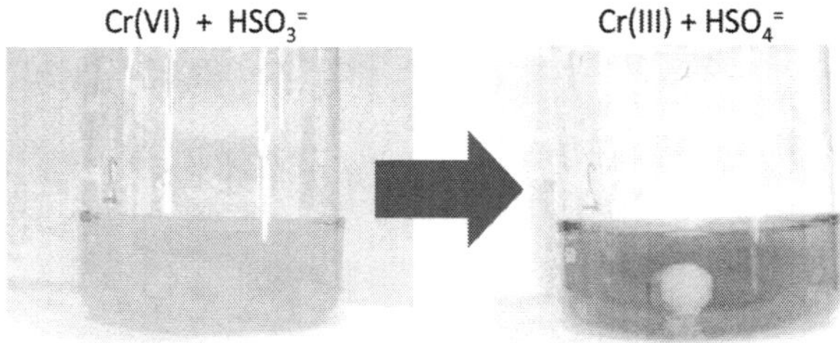

Figure 6. Reduction of Cr(VI) by sodium bisulfite solution. (see color insert)

The experiment was then repeated with sodium sulfite, and again with sodium thiosulfate, both of which produced a slightly higher pH. In these tests nothing happened within an hour, which suggests that pH may again play a role. A strong odor of sulfur compounds was also detected from sulfites, again making these compounds not the best choice for use in a polyethylene plant environment.

Reduction by Iodide

Iodide is another well-known reducing agent for Cr(VI) (*2, 3*). It has even been used as a quantitative measure of Cr(VI) on Phillips polymerization catalysts (*29, 30*). In one test an excess of potassium iodide solution was added to a catalyst slurried in water at neutral pH. No reaction occurred.

In another test a sodium chromate solution was treated with an excess of potassium iodide at neutral pH, and again no reaction was observed. Figure 7 shows this solution at left. However, when a small amount of sulfuric acid was added an immediate color change occurred due to the formation of a deep red-brown iodine solution. Once again, the reduction would only proceed under acidic conditions. Because of this, and the general undesirability of iodine by-product, iodide was again considered a poor choice for application in a PE production plant.

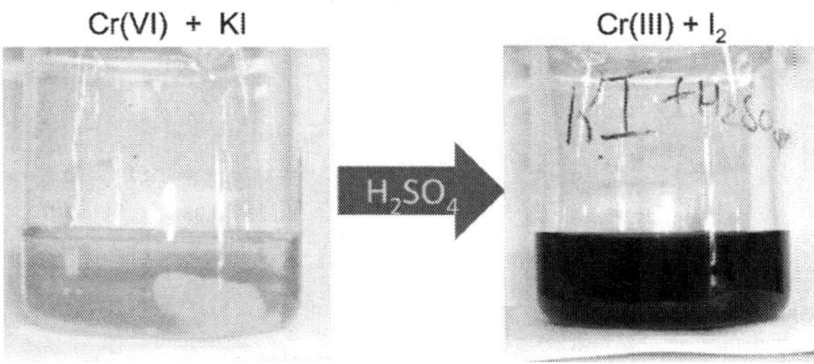

Figure 7. Reduction of Cr(VI) by potassium iodide. (see color insert)

Ascorbic Acid

By far, the most useful reducing agent tested was ascorbate, or vitamin C. Ascorbic acid solution instantly reduced Cr(VI)/silica catalyst to the trivalent form. Although ascorbic acid is weakly acidic, calcium ascorbate and sodium ascorbate were also tested with equal success. Reports by others have also indicated that the reaction between ascorbate and Cr(VI) solutions is quantitative and very fast, even at very low concentrations (*2, 5*). Figure 8 shows the color change caused by addition of ascorbic acid to a catalyst solution at near neutral pH. It has more of a blue color than in other tests, which perhaps suggests chelation between Cr(III) and the oxidized ascorbate ion, or (less likely) reduction to Cr(II).

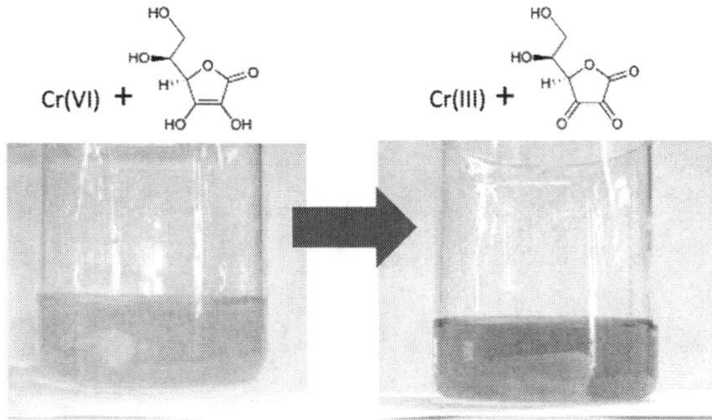

Figure 8. Reduction of chromate by ascorbic acid. (see color insert)

However, in the actual treatment of Cr/silica catalyst spills, it is likely that the catalyst could be in immediate contact with only a limited amount of ascorbate solution. This is because the Cr(VI) is attached to the silica surface within the porous network of the particle. Typically, the catalyst has a pore volume of about 1.6 mL/g, which means that when the ascorbate solution is sprayed onto the catalyst it can adsorb only 1.6 mL/g of liquid. This is the amount that instantly adsorbs by capillarity to neutralize the 1 wt% Cr(VI) in the structure. Any liquid in excess of 1.6 mL/g merely runs off, and cannot contact the Cr(VI) until it migrates out, by diffusion, or the unspent ascorbate diffuses in, both of which probably take place more slowly.

Fortunately ascorbate is highly soluble in water, so that it is possible to make suitable concentrated solutions. Therefore to test its application in an actual catalyst spill test, where only about 1.6 mL/g of liquid is adsorbed, a solution of ascorbate was made that was concentrated enough to neutralize all Cr(VI). It was then charged to a pump-sprayer that held about three gallons of liquid, being originally designed to apply home lawn and garden chemicals. With this device, the liquid could be easily sprayed over dry floor surfaces, shoes, and even thicker piles of catalyst.

To test the solution, about five grams each of three typical Cr(VI) catalyst types were piled onto petrie dishes. Then a brief spray of the solution was applied, but not enough to wet the entire amount. Instead, only a portion in the center was wetted, so that any color change could be distinguished easily. The three catalysts were 1) a simple Cr(VI)/silica, 2) a Cr(VI)/silica-titania, and 3) a Cr(VI)/silica-titania-magnesia. In each case a color change was noted where the liquid was adsorbed.

The first catalyst changed as expected from the original orange color of Cr(VI) to the green color of Cr(III). However, the other two catalysts surprisingly produced a different color change. The second catalyst changed from yellow Cr(VI) to a red color, and the third catalyst changed from a yellow-green color, from the presence of a small amount of Cr(III) accompanying the predominant Cr(VI), also to a red color. Further testing indicated that, despite the red color, these two catalysts were indeed reduced. However, the oxidized ascorbate ion chelated with the titanium in the catalyst to produce the unexpected brilliant red color. Thus, the method was successful in all three tests, which are shown in Figure 9.

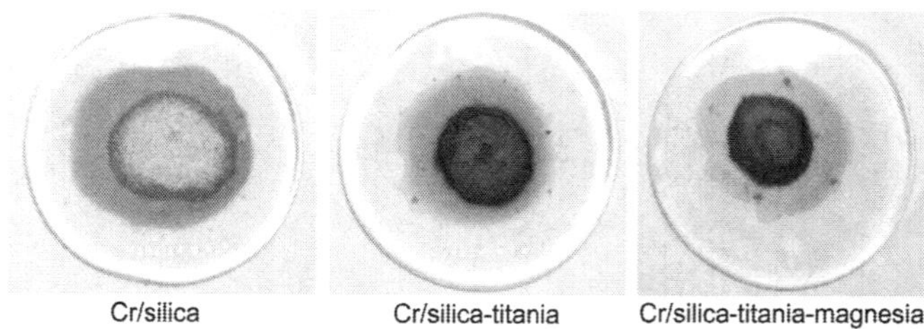

Figure 9. Cr(VI)/silica catalyst is reduced by ascorbate to the green color of Cr(III)/silica, but if titanium is present it then chelates with the oxidized ascorbate to form a red complex. (see color insert)

Ascorbate would then seem to be an ideal answer to contamination in a polyethylene plant. Only one drawback needs to be mentioned. Solutions of ascorbate do react with oxygen in the air over time to lose their potency. We prefer to make up fresh concentrated solutions at least each month.

Figure 10. Diphenyl carbazide indicator turns a deep purple in the presence of Cr(VI). It can be used to detect catalyst dust on surfaces. (see color insert)

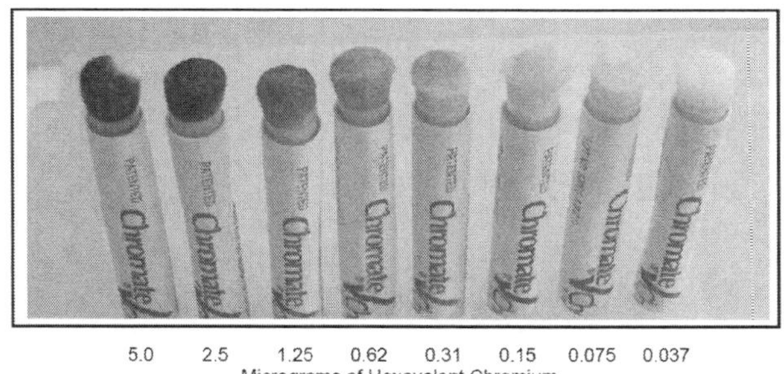

Figure 11. ChromateCheck calibrated test swabs to detect Cr(VI) on a surface. Note: 3 mcg Cr(VI) per swipe would require action. (see color insert)

Detection of Trace Amounts of Catalyst

One additional issue that sometimes confronts plant operators is how to know whether a surface has been contaminated by microscopic amounts of catalyst dust, that can be invisible to the unaided eye. The indicator mentioned earlier, diphenyl carbazide, provides a convenient answer. It reacts with even traces of Cr(VI) to form a deep red-purple color that indicates catalyst is present. A portion of the surface to be inspected is wiped with a cloth containing a weakly acidic solution of diphenyl carbazide. If any color develops upon contact, the surface can be deemed contaminated and then must be sprayed with ascorbate solution to decontaminate it. Figure 10 shows the usual color of diphenyl carbazide on contact with Cr(VI) solution.

One can make up the diphenyl carbazide solution to the desired concentration, and in the preferred container (squeeze bottle, sprayer, wipe cloth, etc) as needed. However, there is another, very convenient, way of using the indicator. A commercial product, called ChromateCheck, is sold as diphenyl carbazide in ampoules with a wick or swab at one end. The ampoule is broken upon use and the swab rubbed over a surface. It turns red-purple if Cr(VI) is detected. In fact, the product is calibrated to give the amount of Cr(VI) detected, by comparing the observed color intensity to an accompanying reference chart. The product is not expensive, has an indefinite shelf life, is easy for even non-technical operators to use, requires no laboratory preparations, and can be discarded after each ampoule is spent. Each kit contains 8 swabs, an instruction sheet and a color guide. The ChromateCheck product is shown in Figure 11.

Conclusions

In contrast to the widely accepted iron (II) salts, ascorbate (vitamin C) appears to be an ideal answer to catalyst contamination in a polyethylene plant. It is a good choice for neutralizing spills, for clean-up of surfaces where there is potential for human contact, e.g. floors, counter tops, shoes, clothes, etc, for neutralizing sump ponds, and just for prophylactic applications (i.e. "just to be sure"). Ascorbate is highly reactive with $Cr(VI)$ catalyst, reducing it quantitatively to $Cr(III)$, even in neutral pH or alkaline solutions, and in very low concentrations. It has a low cost (approximately \$2/kg), and is nontoxic, without leaving any undesirable stain or residue. It is highly soluble and available in several forms. Its only defficiency is its tendency to react with air, so that open solutions should be refreshed about every month.

Acknowledgments

We wish to thank Charles R. Nease of Chevron-Phillips Chemical Co. for calling our attention to ChromateCheck test swabs.

References

1. McDaniel, M. P. In *Advanced Catalysis*; Gates, B. C.; Knoezinger, H.; Jentoft, F. C. Academic Press, Elsevier: 2010; Vol. 53, Chapter 3, pp 123–606.
2. Guertin, J.; Jacobs, J. A.; Avakian, C. P. *The Chromium (VI) Handbook*; CRC Press: 2005.
3. Cushnie, G. C. *Pollution Prevention and Control Technology for Plating Operations*, 2nd ed.; National Center for Manufacturing Sciences: Ann Arbor, MI, 2009; ISBN 978-0-615266-39-8.
4. Westhemeri, F. H. The Mechanisms of Chromic Acid Oxidations. *Chem. Rev.* **1949**, *45*, 419.
5. Xu, X. R.; Li, H. B.; Li, X. Y.; Gu, J. D. *Environ. Toxicol. Chem.* **2005**, *24* (6), 1310–1314.
6. Rivero-Huguet, M.; Marshall, W. D. *Chemosphere* **2009**, *76* (9), 1240–1248.
7. Xu, X. R.; Li, H. B.; Li, X. Y.; Gu, J. D. *Chemosphere* **2004**, *57* (7), 609–613.
8. Gilbert, E. Chromium (VI) Treatment Technologies. http://www.clu-in.org/contaminantfocus/default.focus/sec/chromium_VI/cat/Treatment_Technologies/.
9. Nkhalambayausi-Chirwa, E. M.; Wang, Y. T. *Water Res.* **2001**, *35* (8), 1921–1932.

10. Cook, K. R. *In-Situ Treatment of Soil and Groundwater Contaminated with Chromium*; Technical Resource Guide, EPA 625-R-00-005, 2000.
11. Cieslak-Golonka, M. *Polyhedron* **1989**, *15* (21), 3667–3689.
12. Wielinga, B.; Mizuba, M. M.; Hansel, C. M.; Fendorf, S. *Environ. Sci. Technol.* **2001**, *35* (3), 522–527.
13. Patterson, R. R.; Fendorf, S. *Environ. Sci. Technol.* **1997**, *31*, 2039–2044.
14. Fendorf, S. E.; Zasoski, R. J. *Environ. Sci. Technol.* **1992**, *26* (1), 79–85.
15. Griffin, G. M. U.S. Patent 5256306, issued Oct. 26, 1993.
16. ATSDR Agency for Toxic Substances and Disease Registry. *Public Health Statement Chromium CAS # 7440-47-3*; Division of Toxicology and Human Health Sciences: September, 2012.
17. Jung, Y. J.; Lee, W. *Geophys. Res. Abst.* **2005**, *7*, 04530.
18. Fendorf, S. E.; Li, G. *Environ. Sci. Technol.* **1996**, *30* (5), 1614–1617.
19. Eary, L. E.; Rai, D. *Environ. Sci. Technol.* **1988**, *22* (8), 972–977.
20. Harris, D. C. In *Quantitative Chemical Analysis*, 4th ed.; W. H. Freeman: New York, 1995; ISBN 0-7167-2508-8.
21. Palmer, C. D.; Wittbrodt, P. R. *Environ. Health Perspect.* **1991**, *92*, 25–40.
22. Bartlett, R. J.; Kimble, J. M. *J. Environ. Qual.* **1976**, *5* (4), 383–386.
23. Loyaux-Lawniczak, S.; Lecomte, P.; Ehrhardt, J. *Environ. Sci. Technol.* **2001**, *35*, 1350–1357.
24. Bartlett, R. J.; Kimble, J. M. *J. Environ. Qual.* **1976**, *5*, 379–386.
25. Bloomfield, C.; Pruden, G. *Environ. Pollut. A* **1980**, *23*, 103–114.
26. Palmer, C. D.; Puls, R. W. *Ground Water* **1994**, EPA/540/S-94/505.
27. Eckert, J. M.; Stewart, J. J.; Waite, T. D.; Szymezek, R.; Williams, K. L. *Anal. Chem. Acta* **1990**, *289*, 180–213.
28. Deng, A.; Stone, A. *Environ. Sci. Technol.* **1996**, *30*, 463–472.
29. McDaniel, M. P.; Burwell, R. L. *J. Catal.* **1975**, *36*, 394.
30. McDaniel, M. P.; Burwell, R. L. *J. Catal.* **1975**, *36*, 404.

Chapter 8

Scale Up in Brewing: Factors in Changing Batch Size from 5 Gallons to 15 Barrels

C. J. Archambault,[1,*] W. R. W. Gerds,[2] and A. M. Mills[3]

[1]Par Pharmaceuticals, 870 Parkdale Road, Rochester, Michigan 48307
[2]Lynwood Brewing Concern, 4821 Grovebarton Road, Raleigh, North Carolina, 27613
[3]Cranker's Brewery, 213 S State Street, Big Rapids, Michigan 49307
*E-mail: archamc3@gmail.com

The process is the same for beer brewing as it has been for centuries. The malted grain is milled and its stored starches converted to sugars and alcohol. The equipment however has changed considerably with time and production volume. Due to the massive surge in hobby brewing, the quality and availability of small scale brewing equipment has changed the approach to brewing small batches. The 5 gallon brewing equipment described herein is generalized for simplicity's sake and reflects the equipment used by the brewers at Cranker's Brewery to produce pilot batches.

Ingredients

The list of ingredients to make beer has 4 major components the grain (malt), hops, yeast, and water. Variations in each of these leads to the wide diversity of product currently available in the market place.

Grain

The first step in beer production is the grinding of grain. The goal is not to make a fine powder of the grain, but to slightly crack the grain. This ensures that the starches and enzymes are readily available when doughing in (the mixture of grain and water that will soon become wort or pre-beer) without destroying the grain husk that ensures a proper filter bed. Most beer uses a base of barley malt, however wheat, corn, rice, sorghum, millet, oats and any other starch bearing grain can provide the source for fermentable sugars.

Hops

Hops or *Humulus Lupulus* are a perennial climbing plant that grows in temperate regions throughout the world. Hops provide flavor, aroma, and stability to a beer. The hops contain compounds called alpha acids. When these alpha acids are heated, they isomerize to produce bitter compounds with anti-microbial properties. The hops also contain many oils and flavoring compounds that lend many favors to the finished beer. The concentration of these different hop flavor compounds can lead to many different styles of beer. The differences in hop strain can also lend different flavors to beer. Some common varieties of hops are Cascade, Golding, Hallertauer, Saaz, Nugget and Chinook.

Yeast

Yeast, without it there would be no beer. Yeast is a unicellular fungus that reproduces asexually by budding. There are two primary types of yeast used to produce beer, top fermenting ale yeast (*Saccharomyces cerevisiae*) and bottom fermenting lager yeast (*Saccharomyces pastorianus*) The yeast is responsible for the production of alcohol and many flavor chemicals in the beer. Top fermenting yeast ferment at higher temperatures and produce a larger proportion of esters and phenolic compounds that lend flavor to the beer. Bottom fermenting yeasts ferment for longer periods of time at cooler temperatures and produce less esters then their top fermenting counterparts. Yeast occurs naturally in the air and on the surface or the grain. Over time brewers have separated and cultivated specific yeast strains due to their ability to produce a consistent and desirable product.

Water

Beer is mostly made of water. It has been noted that the quality of the water can have great effect on the finished product. Beer styles the world around have been affected by the base water used in their making. The hardness of the water caused by Calcium, Magnesium, and Bicarbonate can greatly change how the water reacts during the brewing process. Other elements like sulfur and chlorine in the water can cause off flavors in the finished beer.

Control of process water used in brewing is a concern when scaling up. When brewing at the 5 gallon size the brewer can use municipal water supply or purchase bottled water and then add or remove the proper salts to change the hardness of the beer. At the 15 bbl. Batch size (472.5 gallons), source of water is limited. Many breweries source their water form municipal water supplies or local wells and then change the water profile after determining the natural water profile present in the source water.

Equipment

Grain Mills

The equipment needed to grind the grain changes significantly from 5 gallon brewing to production brewing. For anything under 20 lbs. of grain, it is not uncommon for a hobbyist to use hand crank mills with adjustable plates or rollers to control grain cracking. However the hundreds of pounds of grain necessary to produce larger batches of beer requires a larger mill with hopper and augur to move the massive quantities of grain. The larger milling operation is a source of danger because the mills used can easily damage appendages that stray too close. The production of dust during the milling process is hazardous to any that might breathe in the fine particulates. Brewers tend to wear particulate masks to protect from grain dust during the milling process. The dust also carries with it many problems for the brewer related to sanitation, a prime concern. The dust carries with it bacteria and other microorganisms responsible for the spoilage of beer. Many breweries partition the grinding process to ensure that contamination form the grain dust does not occur.

Mash Tun

Inside the mash tun, the grain is mixed with the water in a process called doughing in. The production equipment for 5 gallon batch sizes often include a proportionally sized cooler with a false bottom or a braided line to separate grain from the wort, while keeping the grain at the appropriate temperature for enzymatic starch conversion.The product of enzymatic conversion is wort, a sugary solution that yeast uses as a food source. The production sized equipment must meet the same guide lines yet contain the larger amount of water and grain necessary to produce the wort.

For Cranker's Brewery, the larger production equipment sustains a higher extraction efficiency of 80% - 85%, while the 5 gallon brewing equipment produced extraction efficiency at approximately 75%. Some of this difference in efficiency is the process used to wash the grains. The production brewing equipment uses a constant flow of water on to the grain to wash the sugars through the filter bed (fly sparging). While in a typical 5 gallon pilot brewing system, the wort is rinsed with a single addition of water and then drained through the filter bed (batch sparging). However the sparging process is not the only factor in efficiency for the mashing equipment.

Inside the production brewing mash tun there are rakes or paddles used to thoroughly stir the grain and water mixture. The rakes provide a more homogenous product ensuring that no clumps form in the grain. This effective mixing coupled with the process of fly sparging allows a filter bed of barley husks to form at the bottom of the mash tun. The brewer has a greater control of the sparging process in a 15 bbl. system and is able to ensure that compression of the grain bed does not occur. Compression would result in an ineffective sparge process with a loss of efficiency. This results in better lauterability, meaning the water can more evenly filter through the grain bed to collect the sugars in larger production brewing equipment.

Efficiency

The greater extraction efficiency observed in the production scale brewing equipment provides interesting challenges in scaling up the grain bill. If the brewer directly scales the grain bill by volume of beer it would provide a beer with a stronger malt flavor profile, that contained a higher concentration of sugars, and was darker then the intended beer. With higher efficiency in extract the brewer must scale back the base malt to achieve the desired concentration of sugars often expressed in terms of specific gravity. The greater extraction of flavor and color compounds from specialty malts would also necessitate a decrease in their proportion to the base malt until a balance is found to achieve the correct contribution of each grain.

The Kettle

The process of boiling the wort is not without its own difficulties in scale up. The boiling of the beer serves several functions. First is the sanitization of the sugary wort, this ensures that all of the microbes present in the grain do not cause spoilage and only the yeast added to the wort grows. The second function of boiling is to isomerize the alpha acids from the hops creating bittering compounds that will also reduce the chance of spoilage in the finished beer. Thirdly, the boil serves to remove unwanted grain components. The boiling causes the volatilization of compounds like DMS or Dimethyl Sulfides (responsible for a sweet corn off-flavor). The boil also serves to denature and coagulate proteins that may cause haze in the finished beer.

The major problem associated with the boil and scale up is that while 5 gallon batch size can be brought to boil on an ordinary stove or turkey burner, 15 bbl. production requires a significantly larger source of heat. With 5 gallon brewing, stirring the wort by hand to prevent scorching and burning is possible however when we scale to 15 bbl. we can no longer stir the volume of beer effectively, this is overcome by the use of a pump. The beer is circulated and returned to the kettle at an angle to cause a whirlpool to form.

In a 15 bbl. system the whirlpool effect also effectively produces a cone of collected Trub (coagulated proteins) and hops in the center. The whirlpool then allows the production brewer a higher volume of post boil wort due to improved removal of the post boil detritus (spent hops and protein).

Hop Usage

The use of hops changes with scale up as well. Alpha acid utilization is a function of temperature, boil time, and specific gravity of the wort. The major difference between 5 gallon brewing and 15 bbl. brewing when it comes to hop utilization is the time spent at hop isomerization temperature. In 15bbl brewing the addition of the whirlpool step to separate the boil detritus from the wort results in an extended period between boil and chilling of the beer. The longer the hop compounds stay at high temperatures the greater percentage that are fully converted to their isomerized forms. This means that a 15 bbl. production brewer has to use less hops to achieve that same amount of bitterness in the finished beer. However this also means that a greater amount of the hop volatiles are lost, these hop volatiles are responsible for the flavor and aroma of the beer. During recipe scale up this change in utilization requires the brewer to change the amount of hops added throughout the boil and the times at which the hops are added to ensure that the hop profile matches expectations.

Wort Chiller

Another factor that must be taken into account when scaling up is determining how to cool the resulting wort. The wort must be chilled to at least 100 degrees Fahrenheit before the addition of the yeast however it is preferable to cool to 60-70 degrees Fahrenheit. The 5 gallon batch brewer has several avenues to cool the wort. Some of the most common methods include large amounts of ice or an immersion chiller (a coil of metal tubing in which water flows to exchange heat). Ice is impractical for larger batches and the immersion chiller design is not suited to the large volume of wort nor the amount of thermal energy that must be transferred during cooling of a 15 bbl. batch. The issue is solved by the use of a plate frame heat exchanger. The plate frame design has a larger surface area to exchange the thermal energy and often uses other coolants in addition to water.

The Fermenter

Fermenters commonly come in three different materials and 3 common shapes. The three most common materials used to make fermenters are plastic, glass, and stainless steel. The most common fermenter designs are a straight sided bucket with lid, a carboy, and a conical tank. The majority of 5 gallon brewing equipment consists of plastic buckets, or carboys made of glass or plastic. While almost all fermenters used for 15 bbl. production brewing are conical fermenters made of stainless steel. Conical fermenters used in 15bbl brewing represent a great convenience when it comes to removing yeast, hops, and protein from the finished beer. Because of their shape, the sediment settles into the cone and is easily removed from a purge valve.

Liquid Transfer

In general, the transfer of the wort from mash tun, to kettle, to fermenter, and finally to bottle in 5 gallon batch sizes is achieved by using gravity transfer. While some production breweries are designed to perform on gravity feed, most often production breweries use wort pumps to move the liquid to and from individual pieces of equipment. Wort pumps are also necessary to achieve the aforementioned whirlpool action after boil and during initial heating to avoid scorching.

The pumps are a cause for concern for the safety conscious. Mechanically, if one of the pumps fails or becomes damaged it could taint the final product and cause a great deal of harm to the brewer, especially post boil when the pump is transferring extremely hot liquids, or during the cleaning process described below.

Cleaning

There are two important cleaning processes that are necessary in brewing, the removal of any soils and the sanitization of equipment. Sanitary procedure is extremely important throughout the brewing process to ensure that only the desired culture of yeast proliferates and metabolizes the sugary wort solution.

The scale up from 5 gallon batch brewing must take into account the larger surface area required to clean. For a 5 gallon batch brewer, the use of a scrub brush and oxygen-based cleansers are sufficient to remove brewery soils (proteins, sugars, and sediment). While sanitization is achieved through the use of solutions containing iodine, metabisulfate, chlorine, or starsan (an acid based sanitizer).

Although some hand cleaning is used in 15 bbl. production brewing, the brewer cannot remove the larger quantity of soils by hand as is possible with 5 gallon pilot batches. CIP (Clean in place) systems are installed in the equipment and aid with the cleaning process by using pumps to spray pressurized chemicals into the equipment. In a 15 bbl. system, the removal of soils by the CIP system involves a two part chemical treatment. First the system is washed with an alkaline solution containing sodium hydroxide to remove protein wastes and organic scale (beer stone). The second cycle is an acid mix containing Phosphoric and Nitric acids. These acids remove inorganic scale as well as any organic scale not removed by the caustic wash. The final sanitation process is achieved via a mixture containing hydrogen peroxide, peroxyacetic acid, and acetic acid. The strong oxidative chemicals coupled with the acidic environment ensure that no viable microorganism remain. These chemicals present a higher danger to the brewer then those used in a 5 gallon system. For the production brewer the low pH of the acid mixture and the high pH of the Sodium hydroxide wash respectively present a safety concern because of their ability to contact hazard for skin. These chemicals are cause for concern both in their concentrated form as well as in a dilute solution. The sanitizer also presents a significant safety risk to the brewer because the strong oxidizers present can cause irreparable harm to on contact with the chemicals prior to dilution.

Conclusion

The process of scale up from a 5 gallon batch size to a 15 bbl. batch size is strongly effected by the change in equipment. The larger production equipment engineered especially to produce beer provides the brewer with timesaving alternatives that decreased the time required to brew and increase the efficiency of the entire process. The equipment and its increased efficiency provides the production brewer with a decreased investment raw materials and a higher yield of product. For those with little knowledge of the brewing process, and the factors in effect during scale up, this chapter can provide a basis for understanding the complex processes that occur during brewing. With knowledge, the processes and equipment can then be manipulated to produce a better end product. The knowledge gained from brewing scale up can be applied to other manufacturing processes in an effort to understand and improve the process based on the unique challenges present.

Editors' Biographies

Mary K. Moore

Ms. Mary K. Moore is an Innovation Process Manager in Process and Applications Innovation, Strategic Technology at Eastman Chemical Company in Kingsport, Tennessee. Ms. Moore has extensive experience in designing, constructing, and operating lab, bench, and pilot scale units for separation processes, organic synthesis, organic metallic chemistry, and experimental scale-up. She has a broad range of experience in experimental design and data analysis. She has received seven United States Patents. Ms. Moore graduated in 1991 from Northeast State Technical Community College with an Associate of Applied Science degree in Chemical Technology and was the recipient of the 1991 Outstanding Graduate Award. In 2012, she received the Outstanding Alumna Award. She has served in a number of positions in the American Chemical Society (ACS): Division of Chemical Technicians, Committee on Technician Affairs, Industrial and Engineering Chemistry Division, ACS Career Services, Division Activities Committee, Northeast Tennessee Section of the ACS, and Tennessee Government Affairs Committee. She was honored to be named to the 2009 ACS inaugural class of Fellows.

Elmer B. Ledesma

Dr. Elmer B. Ledesma is an assistant professor in the Department of Chemistry and Physics at the University of St. Thomas in Houston, Texas. His expertise and research specialty is in the pyrolysis, gasification, and combustion of fossil fuels and biomass. Dr. Ledesma conducted his dissertation research on the pyrolysis and combustion of coal volatiles at the Commonwealth Scientific and Industrial Research Organisation (CSIRO) Division of Coal and Energy Technology in North Ryde, Australia, and was awarded his Ph.D. by the University of Sydney. He conducted postdoctoral studies in the Department of Mechanical and Aerospace Engineering at Princeton University and worked as a research associate in the Department of Chemical Engineering at Louisiana State University. Dr. Ledesma serves as a program chair in the ACS Division of Industrial & Engineering Chemistry and is a reviewer for *Energy & Fuels* and *Industrial &Engineering Chemistry Research*. He is a member of the ACS (Division of Energy & Fuels and Division of Industrial & Engineering Chemistry), the International Union of Pure and Applied Chemistry, the Combustion Institute, and the American Institute of Chemical Engineers.

Indexes

Author Index

Subject Index